D1164065

BIOTECHNOLOGY, AGRICULTURE AND FOOD

ORGANISATION FOR ECONOMIC CO-OPERATION AND DEVELOPMENT

TP
248.65
F66
B545
1992

ORGANISATION FOR ECONOMIC CO-OPERATION AND DEVELOPMENT

Pursuant to Article 1 of the Convention signed in Paris on 14th December 1960, and which came into force on 30th September 1961, the Organisation for Economic Co-operation and Development (OECD) shall promote policies designed:

- to achieve the highest sustainable economic growth and employment and a rising standard of living in Member countries, while maintaining financial stability, and thus to contribute to the development of the world economy;
- to contribute to sound economic expansion in Member as well as non-member countries in the process of economic development; and
- to contribute to the expansion of world trade on a multilateral, non-discriminatory basis in accordance with international obligations.

The original Member countries of the OECD are Austria, Belgium, Canada, Denmark, France, Germany, Greece, Iceland, Ireland, Italy, Luxembourg, the Netherlands, Norway, Portugal, Spain, Sweden, Switzerland, Turkey, the United Kingdom and the United States. The following countries became Members subsequently through accession at the dates indicated hereafter: Japan (28th April 1964), Finland (28th January 1969), Australia (7th June 1971) and New Zealand (29th May 1973). The Commission of the European Communities takes part in the work of the OECD (Article 13 of the OECD Convention). Yugoslavia has a special status at OECD (agreement of 28th October 1961).

Publié en français sous le titre :

BIOTECHNOLOGIE,
AGRICULTURE ET ALIMENTATION

© OECD 1992
Applications for permission to reproduce or translate all or part of this
publication should be made to:
Head of Publications Service, OECD
2, rue André-Pascal, 75775 PARIS CEDEX 16, France

26406233

/ 1656

FOREWORD

Biotechnology, Agriculture and Food follows six earlier OECD expert reports issued between 1982 and 1992 that have assessed various topics and policy implications of biotechnology. A logical link leads from the general report, *Biotechnology: Economic and Wider Impacts* (1989) to the present specific one.

The group of experts (further details in list of participants) who wrote the report, are Prof. Sir Barry Cross (Biotechnology, Policy and the Future and Chapter II, Animals), Prof. Richard B. Flavell (Chapter I, Plants and Chapter IV, Non-Food Products), Prof. R. Brian Heap (Chapter II, Animals), Prof. R.J. Roberts (Chapter II, Fish), Prof. Douglas L. Georgala (Chapter III, Food), Drs. Jan Drozd, Willy De Greef and Riccardo Galli (Chapter IV, Non-Food Products), Drs. Bernardo Sorj and John Wilkinson (Chapter V, Strategies of Leading Firms), Prof. Louis Lemkow (Chapter VI, Public Acceptance), R. Stephen Crespi (Chapter VII, Patent Protection), Dr. Volker Beusmann and, from the Secretariat, Dr. Salomon Wald (Chapter VIII, Economic Impacts). The Glossary was provided by Dr. Christina Goodacre-Forsling and Prof. Sir Barry Cross.

A larger expert group, chaired by Prof. Daniel Thomas, met twice to review the report. The first expert seminar took place in Novara, Italy, on 6-8 March 1990; the second, to discuss the final draft, in the Château de Bellinglise, France, on 15-17 May 1991.

The general co-ordination of the work was carried out by Dr. Salomon Wald from the OECD Secretariat, who also wrote the Overview. He was assisted by Dr. Jan Drozd, from the United Kingdom.

The report is unlike some of the previous publications on the subject. Its aim is to review the important facets of agrofood biotechnology in a form that, although comprehensive and scientifically up-to-date, should also be accessible to an interested lay public and to policy makers. It pays particular attention to balance and coherence between the individual scientific chapters. This also explains why the scientific authors have been chosen from the same country so that they were able to meet regularly and work as a team.

Part I reviews the large and growing number of scientific and technological options, Part II the delays and obstacles, on the way to their realisation, which suggest that the diffusion of agrofood biotechnology will be gradual, without major destabilising socio-economic impacts in the OECD area.

The numerous, important issues related to safety assurance and to developing countries are not treated in separate chapters, but are underlying themes, mentioned throughout the report. Safety is already the topic of an other, major OECD activity, which has led to the new publication, *Safety Considerations for Biotechnology – 1992*.

Thanks are due to the industrial companies which co-operated with the interviewers responsible for Chapter V on strategies of leading firms (a project which was co-financed by the OECD Development Center), as well as to the Member countries and organisations which provided support for the report, particularly: FERTEC, Novara, for sponsoring the first expert seminar; Le Biopôle, Amiens, and the region of Picardy for sponsoring the second expert seminar; the Commission of the European Communities (DG XII), for financial support; the Government of the United Kingdom (Laboratory of the Government Chemist, Department of Trade and Industry) for making Dr. J. Drozd available to the OECD; the Government of Germany (Federal Ministries for Research and Technology and for Food, Agriculture and Forestry) for making Dr. V. Beusmann available to the OECD.

The Committee for Scientific and Technological Policy discussed the report as well as the overview and policy conclusions at its 56th Session on 24 October 1991. The delegates expressed satisfaction with the professional quality and timeliness of the report. Two chapters in particular, "Public Acceptance" and "Patent Protection", attracted comments. It was felt that the government approach to public acceptance did not need to be defensive. For example, the most recent opinion survey on biotechnology ("Eurobarometer" 1991) had given encouraging results. Public acceptance would ultimately be conditioned by confidence in regulatory policies. In this context, delegates also stressed the importance of the safety and environmental impact issues. Regarding intellectual property rights, delegates suggested taking a broad view, beyond the existing legal framework, for example by considering the links between patent protection and ethical questions.

The Committee also raised various other issues. Among them were the definition of biotechnology, Third World problems, training and human resource requirements and the potential contribution of further technology assessment studies to identify appropriate R&D and innovation strategies. The policy conclusions of the experts led to the comment that they did not reflect the full complexity of the report. However, it was also argued that recommendations had to be simple if they were expected to have a policy impact.

The final text of the report takes this discussion into account. The Committee recommended its publication on the responsibility of the Secretary General. The report does not necessarily reflect the views of the OECD or of its Member governments.

TABLE OF CONTENTS

OVERVIEW AND POLICY CONCLUSIONS 11

BIOTECHNOLOGY, POLICY AND THE FUTURE 29

Part One

SCIENTIFIC AND TECHNOLOGICAL OPTIONS

Chapter I

PLANTS

1.	Introduction	33
2.	Diagnostics	35
	a) For molecules and diseases of concern	35
	b) For crop quality and industrial use	37
3.	New Genes and Variants of Existing Genes for Crops	37
	a) Sources of genes	38
	b) Genetic modification: benefits and risks	43
4.	Improving Plant Breeding Programmes	44
	a) Rapid identification of valuable genes	44
	b) New methods for hybrid seed production	45
	c) Plant propagation and tissue culture	46
5.	Improving Plant Production	47
	a) Herbicide tolerance	48
	b) Plant morphology	48
	c) Stresses	49
	d) Nutrient utilisation	50
	e) Disease	50
	f) Biopesticides	53
6.	Improving Plant Quality	54
7.	Constraints on Biotechnology Applications	56
	a) Property rights	56
	b) Public perception	56
	c) Scientific limitations	56
8.	Prospects for Commercial Exploitation	56
9.	Global Aspects of Plant Biotechnology	59
10.	Further Reading	60

Chapter II

ANIMALS INCLUDING FISH

1. Introduction 61
2. Animal Health 62
   a) Diagnosis 62
   b) New vaccines 64
   c) Other therapeutic agents 65
   d) Animal and human diseases 66
3. Animal Production 66
   a) Animal growth and lactation 66
   b) Nutrition 69
4. Animal Reproduction 70
   a) Embryo multiplication 70
   b) Gene transfer 72
   c) Genome analysis 75
5. Fish 75
   a) Aquaculture 75
   b) Reproduction 76
   c) Transgenesis 78
   d) Disease 80
6. The Future and Public Attitudes 80
7. Further Reading 84

Chapter III

FOOD

1. Introduction 87
2. Food Biotechnology – Options 88
3. Food Biotechnology – Illustrative Examples 90
   a) Food enzyme technology 90
   b) Protein engineering of enzymes 91
   c) Biopreservation 92
   d) Food safety rapid contaminant tests 93
   e) Novel biotechnology foods 94
   f) Plant cell culture 94
   g) Microalgal culture 95
4. Consumer Protection 96
5. Further Reading 97

Chapter IV

NON-FOOD PRODUCTS FROM AGRICULTURE

1. Introduction 99
2. Lignocellulose and Energy 102
3. Starches and other Polysaccharides 106
4. Hydrocarbons, Oils and Fats 107
5. Natural Fibres and Structural Materials 107
6. Pharmaceuticals, Agrochemicals and Fragrances 108
7. Ornamental Plants and Tobacco 110
8. Novel Industrial Products from Transgenic Plants and Animals 110
9. Miscellaneous Products and Processes 111
10. Agricultural Waste Treatment and Biosensors 112
11. Further Reading 112

Part Two

INDUSTRIAL AND SOCIO-ECONOMIC ISSUES

Chapter V

STRATEGIES OF LEADING FIRMS IN AGROFOOD

1. Introduction 117
2. Interview Sample 118
3. Summary of the Major Conclusions 119
4. Upstream Markets: Plants 121
   a) Background 121
   b) Classical breeding vs. plant biotechnology 121
   c) Innovation timetable 122
   d) Hybrid technology 123
   e) Herbicides 123
   f) Bio-insecticides 124
   g) Marketing strategies 126
5. Upstream Markets: Animals 129
   a) Background 129
   b) Red meat 131
   c) Poultry and fish 131
   d) Growth hormone 132
6. Downstream: Intermediates and Primary Processing 133
   a) Enzymes and raw material substitution 133
   b) Primary processing industries 133
   c) Market strategy for enzymes 134
   d) Yeasts and flavours 135
   e) Plant cell culture and developing country resource bases 135
7. Downstream: Final Food Products 136
   a) Importance of biotechnology 136
   b) Quality and biotechnology 137
   c) Final foods, biotechnology and raw materials 137
   d) R&D strategy 138
8. Leading Firms and Major Economic Blocs: Globalisation 138
9. Perspectives for Developing Countries 141
   a) Strategies for raw materials 141
   b) Biotechnology transfer 141
   c) Institutional preconditions 142
10. Conclusions and Policy Implications 143
11. Company Reports Consulted 144
12. Further Reading 144

Chapter VI

PUBLIC ACCEPTANCE

1. Introduction 145
2. Quantitative Studies 146
   a) The United States 147
   b) Europe 147
3. Qualitative Studies 149
4. The Context of Public Attitudes 151
5. Points to Consider 152
6. Further Reading 153

Chapter VII

PATENT PROTECTION
1. General Considerations 155
2. Agriculture and Patents 156
   a) Existing applications 156
   b) The problem areas 157
3. International Practice 157
   a) Legal policies 157
   b) US law 157
   c) Japanese law 158
   d) European law 158
4. Legal Protection for Plants in Europe 158
   a) Plant breeders' rights 158
   b) The response of the plant breeders 159
   c) The views of patent circles 160
5. Patenting Plant Genes 161
6. Patents for Animals 161
7. Patents for Foods 162
8. Opposition to Patents in Biotechnology 162
9. Further Reading 163

Chapter VIII

ECONOMIC IMPACTS
1. Introduction 165
   a) The framework for an economic impact assessment 165
   b) The changing role of agriculture and the food industry in OECD countries 166
   c) Subsidies and price policies at the crossroads: the GATT negotiations 1991 169
   d) Biotechnology and long-term global food supply 170
2. Impacts on Production and Market Balances 171
   a) Quantitative results for the USA (OTA, 1986) 172
   b) The impact of biotechnology on agriculture in the European Community to the year 2005 (Neville-Rolfe *et al.*) 175
   c) Discussion and conclusions 178
3. Impacts on Agricultural Input and Output Prices 179
   a) General effect of technical progress on agricultural input and output prices 179
   b) Specific effects of new biotechnologies on agricultural input and output prices 180
4. Impacts on Farm Structure 182
   a) General framework 182
   b) Impacts on the structure of American agriculture (OTA, 1986) 182
5. Impacts on International Competitiveness and Trade 184
   a) Technological substitution potential in OECD countries for developing countries' crops 186
   b) Main findings on trade impacts in an OECD report of 1989 186
   c) Trade shifts as a result of trade policies 188
   d) Access to biotechnology 189
6. Impacts on Employment 189
7. A Case Study: Economic Impacts of bST 191
   a) Impacts at the farm level 193

	b)	Impacts on structural change	194
	c)	Impacts at national level	195
	d)	Consumer response	196
	e)	Conclusions and issues	197
8.		Summary	197
9.		Further Reading	199

LIST OF PARTICIPANTS 203

GLOSSARY 207

LIST OF FIGURES

1.	A scheme illustrating the production and use of plants	34
2.	The breeding cycle showing conventional methods of gene transfer and evaluation and recently exploited molecular biological methods	39
3.	Scheme for modification of plants by insertion of a single gene to create insect resistant plants	41
4.	Biotechnology impacts on the milk and dairy chain	192

LIST OF TABLES

1.	Useful diagnostics from biotechnology	36
2.	Sources of new genes for plant improvement	40
3.	Crops into which single purified genes have been introduced	42
4.	Species where detailed genetic maps to aid breeding have been created by biotechnological methods	45
5.	Plant characters frequently in need of improvement by exploitation of new genes in breeding programmes	47
6.	Global estimates of losses due to plant diseases, 1987	51
7.	Biological agents with pest control activities	53
8.	Recent biotechnological achievements for potato breeding	57
9.	Impact of plant biotechnology in developed and developing countries	58
10.	Comparison of gains in milk yield per cow for different dairy technologies	61
11.	Examples of diagnostic tests used to monitor fertility and launched between 1984-89	63
12.	Development of a commercial fowl typhoid vaccine from biotechnology	65
13.	Impact of bovine somatotropin (bST) use on animal numbers, feed requirements and waste production of dairy cows to achieve 1988 US milk production	67
14.	Summary of progress of biotechnology-based somatotropins	68
15.	Development of transgenic animals	73
16.	Summary of major research opportunities and potential exploitation	82
17.	Food biotechnology developments	87
18.	Plant tissue cultures developed for industrial application	95
19.	Microalgae and some countries where they are used as food	96
20.	Some options for the use of agricultural systems	100
21.	Approximate annual world production of some agricultural products	101
22.	Comparison of scale of energy production as biomass with energy depletion as fossil fuel	103
23.	Leading firms interviewed according to type and location	119

24. R&D expenditure of a selection of seed and agrochemical firms 122
25. Probable evolution of agrobiotechnology 122
26. World market for crop protection in 1988 124
27. Major core crop markets, 1989 127
28. Industry research in processing improvement 128
29. World animal health market 130
30. US study on acceptability of genetic engineering, 1987 146
31. EC study on genetic research, 1979 147
32. Employment in the agrofood sector in France, 1984 167
33. Consumer expenditures on food, semi-luxury food and tobacco in per cent of
    total household expenditures, EC, 1986 168
34. Historical and projected rates of annual growth in US crop yields 172
35. US animal production efficiencies in 1982 and estimated annual growth rates . . . 173
36. Projections of US crop production 174
37. Projections of US animal production 175
38. Estimated impacts of various new biotechnologies on different crop productions
    to the year 2005 in the European Community 176
39. Estimated impacts of various new biotechnologies on different livestock
    productions to the year 2005 in the European Community 177
40. Trade in farm products 185

OVERVIEW AND POLICY CONCLUSIONS

1. Biotechnology in Agriculture and Food: A Scientific Revolution

When one looks back on the 1980s, it is apparent that the scientific developments underpinning agrofood biotechnologies have been extraordinary both in speed and scope. They are comprehensively reviewed in this report "Biotechnology, Agriculture and Food". New discoveries have been made faster than anyone expected in 1980, and the number and variety of organisms involved are expanding rapidly.

Biotechnology applied to micro-organisms, plants and animals for food and non-food products is more than just a new productivity-increasing or problem-solving technology. Biotechnology is dramatically changing the ways in which scientists grasp the structure and function of biological systems. The knowledge revolution thus brought about will have scientific, ethical and social consequences far beyond the economic effects on agriculture and food. Biotechnology may offer unprecedented possibilities for understanding the critical biological factors of the global ecosystem, and perhaps, a better capacity to anticipate and reduce threats to this ecosystem. This alone would be sufficient reason to continue and even increase the funding of fundamental research in agrofood biotechnologies.

Of course, there has been progress in all fields of science and technology with an impact on agricultural productivity, and it is important to link advances in biotechnology with advances in complementary technologies such as process engineering or sensors in order to realise the full benefits of biotechnology. Nevertheless, the relative importance of the biological, and particularly the genetic, sciences with respect to all other scientific and technological developments affecting agriculture, seems to be steadily increasing.

However, it is a long way from a knowledge revolution to an agricultural revolution. The multiplicity and variety of scientific and technological developments may change the fundamentals of plant and animal breeding and of food production in major ways, but whether they will do so in less than 20 or 30 years depends on numerous factors, many of which lie outside the domain of science and technology: economics, legal and safety constraints, public attitudes, industrial and government policies. In agricultural biotechnology, the road from scientific to economic change could be as long, and the obstacles on the way as numerous, as in any other sector of technology.

Safety assurance issues currently dominate thinking on the problems and constraints arising from biotechnology. Hence, all chapters of the OECD expert report have been written against the background of these issues. If safety in agrofood biotechnology has not been made the topic of a special chapter, it is because it is already the subject of the major OECD activity "Safety in Biotechnology".

Similarly, the beneficial contributions that agrofood biotechnologies can make to the Third World have not been treated in a special chapter but are mentioned throughout the report.

The following examples will point to the advancing frontiers of research and innovation. They illustrate the growing versatility of the new technology and some of the options which may open up in the next ten or more years. A review of industrial strategies and of economic issues and impacts will complement this presentation.

2. Plants

New biotechnology can affect every stage of plant life:

a) Diagnostics

Rapid biotechnology tests for contamination by crop disease organisms and for seed and crop quality controls allow for safer and more efficient crop breeding and trade and will greatly facilitate the application of legally defined standards.

b) New genes for crops

Crop germplasm improvement by the addition of new genes has been the goal of plant breeding since the beginning of agriculture. New efficient genetic modification methods could aim at increasing plant performance and plant resistance to virus and other diseases, as well as to drought, salt, cold, heat, etc. They could also enlarge the land resource basis available for agriculture. Genetic modification might become the most important contribution of biotechnology to plants. From 1982, when the first single gene was successfully transferred, progress has been rapid; several dozen plants have since been modified in the laboratory.

c) Improving plant breeding

In vitro and other biotechnologies help to reduce the time-consuming and expensive process of producing, growing and evaluating large numbers of plants. Included are molecular genetics for rapid identification of valuable genes, new methods for hybrid seed production, and plant propagation and tissue culture.

d) Improving plant production

Crop performance in the field, defined as yield, depends upon numerous factors, including environment, soil type, agronomy, external factors such as pests and disease and the plant properties themselves. Genetic modification of plants or micro-organisms can modify these factors, leading, for example, to better plant morphology, stress resistance, and biological fertilisation, as well as pest and disease control, which would reduce chemical inputs into agriculture.

e) *Improving plant quality*

One of the targets of biotechnology is to improve plant quality so that the harvested parts of the plant are better suited to their end use as foods or non-food products. Progress in this field will tend to blur the borderlines between plant- and food-biotechnologies.

3. Animals

Animal biotechnologies, ranked below by degree of readiness for commercial application, can improve agricultural productivity in many ways.

a) *Improving animal health*

Biotechnology can make a dramatic impact in controlling animal diseases, which are a source of great suffering and economic loss. Examples include new diagnostic agents, molecular vaccines, novel therapeutic agents such as new antibiotics, and transgenic germline modifications to increase natural resistance.

b) *Improving animal growth and lactation*

Production traits such as milk yield and carcass leanness can be enhanced by administration of pituitary hormones made by recombinant DNA technology without harmful effects. The technology has been successful in cattle, sheep and pigs.

c) *Embryo multiplication*

The culture of identical embryos in the laboratory, and procedures for generating embryo banks and multiplication of superior genomes, are well advanced.

d) *Animal feeds*

Genetic modification of micro-organisms concerned in silage production and ruminant digestion offers the prospect of new sources of feed and improved nutrition. Single-cell protein was an early example, based on more traditional technology.

e) *Transgenesis*

Because farm animals have been bred for a restricted number of traits, a vast number of animal genes of potential value have not been identified or exploited. Transgenic technology attempts to improve the usefulness and productivity of animals. This technology has now reached a stage where useful new traits can be introduced into a germline of mammals, fish and birds. Genetic modification is also pursued as a method for having animals produce pharmaceutically important proteins for humans use.

f) *Fish*

Compared to farm animals, fish are still relatively wild, with little breeding history. This, together with the large number of species available, offers a particularly large scope for improvement. Application of biotechnology to fish farming shows promise of great enhancement of high-quality protein supply from aquaculture, through significantly increased efficiency of production and reduction of diseases.

4.   Food

Food biotechnology offers a very wide range of options for improved quality, nutrition, safety and preservation of foods, but no single biotechnology advance will have a dramatic impact across the food processing industry, as the industry is too vast and varied. However, in some cases biotechnology food processing already can change the fundamentals of particular food sectors. Economics, customer acceptance and regulatory requirements rather than scientific hurdles will often dictate the range and speed of food biotechnology applications. Progress will differ from country to country depending on the nature of their food products and the readiness of consumers to accept the benefits of new biotechnologies.

The variety of actual and possible developments is illustrated by the following non-comprehensive list:

a)   *Rapid food safety contamination tests*

Biotechnology (monoclonal antibodies, DNA and RNA probes) is radically changing the type of test methods available to identify, control, reduce or remove contaminants that endanger public health. This will strengthen food enforcement authorities and also assist food processors in rapid routine control of quality and safety.

b)   *Food enzymes*

The many uses which enzymes have in the food and drinks industry for processing, and for the production of flavours and ingredients, will be enhanced by several scientific and technical advances, particularly enzyme bioreactors, enzyme encapsulation and protein engineering leading to new or improved enzymes or novel biocatalysts.

c)   *Biopreservation*

Biotechnology will offer the possibility of biological systems or modified food micro-organisms that can provide preservation capability in a food, without the use of synthetic chemical preservatives.

d)   *Novel biotechnology foods*

Biotechnology offers a means for creating a completely novel human-grade food with a healthful image from lower value materials, such as meat-like protein foods from fungal fermentation (e.g. the myco-protein development).

e)   *Plant cell culture*

A variety of food ingredients and nutrients can be produced by the mass culture of free plant cells through the technique of plant cell culture.

f) Microalgal culture

Microalgae (seaweeds), which have been the basis of traditional human food in a number of countries, can also be grown to produce a wide range of useful compounds, some of which are of importance as human food ingredients or as specialised nutrients.

5. Non-food Uses of Agricultural Crops

Non-food uses of agricultural crops, which already reach 20 per cent or more of agricultural production in many OECD countries, have great economic potential which new biotechnology is enhancing. Economic and environmental parameters, such as the oil price and food price support, regulations for genetically modified plants, and public preference for biodegradable or "natural" products will influence the application of biotechnology to non-food uses of agricultural raw materials.

A variety of examples testify to the growing importance of non-food uses:

a) Wood and energy

Among various technologies to turn wood or other biomass into energy, biotechnology converts biomass into liquid and gaseous fuels through fermentation.

However, as feasibility studies indicate that electricity generation from biomass by gasification may already be competitive and will produce less net carbon dioxide than generation from fossil fuels, a greater energy contribution by biotechnology could come from increased or cheaper biomass production. This would also facilitate the manufacture of numerous other products derived from wood (pulp, substitutes for chemicals, structural materials).

b) Starches and other polysaccharides

While more than half of all starch is used for food and feed, the amount of starch processed into industrial products (paper, textiles, adhesives) is likely to increase, due to new biotechnologies.

c) Hydrocarbons, oils and fats

Natural rubber and oil-bearing crops (used in soaps, detergents, paints, lubricants, etc.) have recently received more attention.

d) Other plants, plant products, animal materials

Ornamental plants, for which there is a rapidly expanding market, lawn grasses, tobacco, cotton and special use plant fibres are open to new biotechnological approaches. Plant products such as some pharmaceuticals, agrochemicals, flavours and fragrances, and animal products such as leather, wool, hair and silk, could also be enhanced.

e) Novel industrial products from transgenic plants and animals

Novel proteins produced by genetically modified plants or animals are a very active area of research (see above).

f)   *Waste treatment*

Residues, wastes and effluents from agriculture and food production – major sources of pollution – can be treated by new and improved biotechnologies, sometimes leading to useful by-products.

Non-food uses benefit from many of the scientific advances mentioned, particularly in the plant sector. Biotechnology can intervene in numerous ways, for example through:

- breeding, cloning and genetic modification of trees to accelerate wood production, and to improve quality;
- improved enzyme-technologies for better product-extraction and conversion (e.g. from starch, oils, fats), and for the manufacture of agrochemicals with greater efficiency and smaller environmental impact;
- micropropagation techniques to accelerate propagation and multiply disease-free stocks;
- genetic modification to generate plants producing new, improved or less polluting substances (oils, pharmaceuticals, agrochemicals, proteins);
- biosensors to detect and monitor pollutants from the agrofood sector.

6.   A Comparative Perspective on Current Trends

The pace of scientific and technological progress has not been the same in the plant, animal, food and non-food sectors. This explains why – aside from the economic and other external constraints – the number of new biotechnologies on the market, and the time scale for commercial exploitation, vary among the four sectors.

Of the four, it is the *food sector* which first saw the arrival of new biotechnological innovations (some of them in the mid-1970s), and where, until recently, more new biotechnologies were in commercial operation than in any of the other three sectors. One reason for this is that "traditional" biotechnologies of fermentation and enzyme application were already established in several food and drink sectors. Modern processing using these biotechnologies was under development and in use and was accepted in consumer and regulatory terms before the molecular biology and "genetic" revolution. The subsequent increase in biotechnology opportunities arising from new "genetic" advances are, of course, now much influenced by the profound transformation of food markets, particularly in respect to consumer assurance, food safety, and regulatory requirements concerning food composition and labelling.

Among the new food biotechnologies currently in commercial operation, one may mention:

- new biotechnology testing methods;
- enzyme- and bio-conversion of starch to sweet products;
- food flavours and enhancers;
- fruit juice processing;
- amino acids and other special nutrients;
- novel structured foods from fermentation;
- cheese enzymes, lactose-free dairy products;
- yeast hybrids.

During the 1980s, the *animal sector* seems to have made faster scientific and technological progress than had earlier been expected. In the 1970s and early 1980s, the obstacles and time scales, particularly in genetics, may have been somewhat over-estimated. Also, the growing interface between animal and human health research has helped both to make progress. Simultaneously, environmental and safety concerns may have caused less delay in animal than in other biotechnology research.

The first new animal biotechnologies – new diagnostics – arrived on the market as early as some of the new food biotechnologies, that is in the late 1970s. Presently in commercial exploitation are:

- animal diagnostics;
- new vaccines and therapeutics;
- *in vitro* fertilisation embryos;
- pituitary (growth) hormone to increase milk yield or growth;
- animal feeds and feed additives.

No animal modified by rDNA-techniques is currently (May 1991) used in commercial agriculture. The only genetically modified animals on the market are laboratory animals used as human disease models (e.g. an "onco-mouse" for cancer research).

In the *plant sector,* a few new biotechnologies are already being exploited for commercial gain, for example:

- plant diagnostics;
- microbial insecticides;
- tissue culture techniques;
- micropropagation techniques;
- genetic mapping techniques.

No new plants, and (except for one case in Australia where approval was granted in 1991) virtually no new micro-organisms genetically modified by r-DNA or other techniques have yet been released into commercial agriculture (October 1991).

In the *sector of non-food uses* of agricultural production, several biotechnology-based products or processes are commercially available:

- conversion of biomass to energy via anaerobic fermentation to ethanol or methane gas;
- selective breeding and propagation techniques for trees and ornamental plants;
- plant cell culture techniques;
- enzyme technologies for product extraction and processing (from starch, vegetable oils, etc.) and for the manufacturing of agrochemicals.

Several of these or similar technologies belong also to the food or plant sectors. Although a genetically modified yeast for baking use was cleared in the United Kingdom in 1990 and cell fusion procedures have been used for yeast improvement elsewhere, no genetically modified products appear to be in commercial production (May 1991).

While there are large differences among the four sectors, there are clearly also a few strong common features. The most striking one is that the earliest, and in some sectors the most widespread and promising, use of new biotechnology focuses on improving health, particularly through new diagnostics, and this both in humans (control of food contamination) and in plants and animals (disease diagnostics and prevention). Another common feature is that genetic technologies, both in food production and in modifying plants or animals for food, are not yet commercial (May 1991). Finally, one of the common aims of many biotechnologies is to reduce the input of chemicals into agricultural and food production.

7. Looking Ahead

The technologies currently (1991) on the market are only a part, and not a very large one, of what is technically already possible. They are an even smaller fraction of the numerous options which could be made available in the next ten to fifteen years.

In some sectors, agricultural biotechnology is on the verge of a quantity and quality jump in innovation, but the technological potential will become a reality only if safety, public acceptance and industrial problems on the way to further commercialisation can be solved during the 1990s and if market forces support the new technologies.

For reasons of consumer acceptance, the *food sector* is not likely to see any revolutionary developments in the immediate future. Some products which are scientifically and technically ready will be commercialised in the 1990-1995 period. After the genetically modified baker's yeast already approved, other new foods for specific nutrition needs, food colours and ingredients, and plant cell and micro-algal cultures for high value ingredients will follow.

In the medium term (1995-2000), a number of additional products that are scientifically and technically feasible are likely to enter the market, particularly genetically modified food bacteria for flavour and quality, modified food enzymes, novel biocatalysts for food processing, bioconversion technology for unusual food environments, and increasing numbers of rapid biotests and biosensors for many food contaminants.

Technological change could be more radical in the *plant sector*. The first transgenic potato plants, cotton, oil-seed rape, tomatoes, tobacco and soybean have been tested in small-scale field trials, and there are commercially valuable transgenic plants of those species that are resistant to specific herbicides, viruses, insects; there are also plants with improved quality traits. Other plant species have been subjected to field trials in non-OECD countries. Technically, some transgenic plants could be on the market by 1995. However, delays due to safety requirements, public concerns and the need to carry out more R&D on genes make it likely that genetically modified plants in large numbers and for a wide range of countries will not be on the market before 2000, and will begin to make a major impact only from 2005 on. Other recent plant biotechnologies will be widely adopted during this period.

In the *animal sector,* too, biotechnology could already bring about noticeable modifications of agriculture in OECD countries by 2000 if safety assurance, public acceptance, industrial and agricultural policy problems can be overcome. Before 1995, numerous new diagnostic kits for the detection of diseases and for fertility and sperm fertilizing ability tests, etc., will be on the market, as will new therapeutic drugs and vaccines (e.g. against parasitic diseases). More new feed additives, including amino acids, antibiotics and rumen micro-organisms will be commercialised, and the first transgenic fish may arrive on the market.

The period from 1995 to 2000 may see the commercial development of pigs and cattle with rapid growth potential, improved carcass quality and for cattle, increased milk production, due to the application of pituitary growth hormone. Also, transfer of commercially valuable genes in animals might allow for multiplication of genetically superior, e.g. disease-resistant, animals.

The *non-food sector,* rather like the food sector, is unlikely to experience any radical changes, except if there is a large increase in the price of oil, or if carbon taxes on fossil fuels are introduced. Genetically modified crops for the production of existing or new products (fine chemicals, pharmaceuticals, enzymes, etc.) could become a commercial reality, but probably not before the year 2000.

8. Bottlenecks and Priorities in Science and Technology

Repeated references to external constraints which delay development must not blind one to the many fundamental difficulties in agro-biotechnology that are of a scientific or technological nature. Limited knowledge and insufficient R&D funding are still among the most important stumbling blocks to be overcome during the next decade. Only some of these are highlighted here.

In the *plant sector, food and non-food crops* included, there is a need for substantial progress in basic plant physiology, biochemistry and in multi-gene transfer across the species barrier. There is a serious shortage of novel valuable genes for crop improvement, as few genes are well characterised and available. Scientists have learned to add genes, but in spite of great progress in 1991, more research is necessary to replace them effectively.

Many crops and trees are genetically not well characterised. The number of crops that can be genetically modified is small. Wheat, beans, cassava, rice and maize are among the major food crops where more progress in genetic modification is required. Research on disease-resistance genes is a particular priority.

Another obstacle lies in the difficulty of regenerating whole plants from genetically modified cells. Comparing food to non-food crops, it appears that R&D attention has focused on the former, at the expense of the latter (with the partial exception of cotton and tobacco), an imbalance which needs to be corrected.

Apart from plant genetics, the genetic modification of micro-organisms for agricultural use is only at its beginning and few real successes have occurred so far in this important sector.

Beyond the immediate plant sector, our understanding of the area of ecology is still too limited, and much more progress is necessary.

In the *animal sector,* just as in the plant sector, only a few useful genes have been identified. Much work will be necessary, for example, to identify disease-resistance genes, to discover how and in which body tissue a gene is switched on or off, and to locate the position of genes on the chromosomes of the animal genome.

Current methods of producing transgenic animals are still very difficult. Moreover, the creation of multiple copies of a particular adult animal remains a fundamental unsolved problem of biology.

In the *food-sector* too, biotechnology suffers from scientific and technological weaknesses, due to a shortage of fundamental research. A better understanding of food molecular structure and function is necessary in general, while in specific technical

19

applications, yield, purity of processing, product excretion by plant cells must be improved, to mention only a few.

To define R&D priorities across the entire spectrum of agricultural biotechnologies will be difficult. Much will depend upon wider policy goals. Biotechnology can be used to promote diverse, even contrasting aims. If the aim is to quickly increase protein supply of high nutritional value for both the developing and the industrialised world, a first effort might be to increase biotechnology R&D for fish. If the aim is to replace chemical herbicides, pesticides and fertilizer, increased support for specific plant and microbial biotechnology research will be required. If the aim is to foster the economic adjustment of agriculture in the OECD area, non-food uses of agricultural crops could be promoted.

9. Industrial Strategies and Constraints

Translating the potential of agricultural and food biotechnologies into products and processes is first and foremost an industrial problem. A survey of the agrofood industry in Europe, the USA and Japan (1990), carried out for this report, indicates that agrofood biotechnology is increasingly dominated, directly or indirectly, by the leading multinational corporations which have a global view both of resource inputs and of markets. Their activity, however, does not reduce the opportunities for the numerous smaller successful biotechnology companies in specialised niche markets, but could even increase them.

The economic viability and diffusion of agrofood biotechnologies depend upon their adoption by the leading industrial actors in the agro-food chain, and upon their successful integration into competitive industrial strategies. This dependence is reflected in a number of particular issues which are relevant to all levels and sectors of the agrofood chain.

a) Particular issues

High levels of uncertainty

Biotechnological innovation in agrofood sectors involves high levels of uncertainty. Partly, this is due to the uncertainties of consumer response, of the safety debate and of intellectual property protection, and partly it is linked to the availability of numerous, often competitive scientific and technological options. Thus, the impressive number of scientific developments mentioned above must be integrated into a complex industrial decision framework.

Quality products and processes

In the OECD area, agrofood markets are dominated by two tendencies: stagnation of traditional outlets, and a profound restructuring of demand which is coupling food to new (and sometimes contradictory) medical, nutritional, ecological, safety or convenience demands. As a result, innovation strategies in agrofood biotechnology tend to favour quality products and processes over cost-cutting applications. Even if cost-cutting remains an important goal in various sectors (seeds, feeds, sweeteners), companies are often shifting away from innovation strategies which can only be justified by cheaper costs, in favour of those which carry a higher quality label, however defined.

New patterns of co-operation

The overriding concern with consumer acceptance and increasing emphasis on quality are leading to new patterns of co-operation among the economic actors at different levels of the agrofood chain. The successful introduction of biotechnology innovations at any level of the agrofood chain must be negotiated with downstream producers and consumers. This includes the difficult question of appropriation and distribution of value added among the different actors of the agrofood system, in order to recuperate innovation costs. Thus, biotechnology has become a dynamic force welding together the various levels of the agrofood chain.

Key role of end-users

The need to reduce uncertainties, to emphasise quality and to foster new forms of co-operation for sharing risks and benefits, all depend on greater responsiveness to end-user demand, whether the end-user is another firm, or the final consumer.

b) Sectors

Plants

The main focus of attention in this sector has been the reorganisation of the seed market, leading to greater integration with the agrochemical sector.

Innovation strategies emphasize hybrid plants and crop protection biotechnologies, such as herbicide and virus resistance in plants, bioinsecticides and fungicides.

Among the marketing strategies for new products, the traditional gene technology supplier option has become vulnerable and is giving way to the strategy of controlling seed markets, or, more importantly, the strategy of moving further downstream into crop output markets, in order to capture the industrial value added.

Animals

Biotechnologies are being developed at all levels of the animal agrofood chain (forage crops, feeds, digestive processes, health, growth, reproduction), with diagnostics and vaccines becoming a major area for small and medium-sized companies. The uneven degree of industrialisation of the main protein sources (beef, poultry, pork, fish) expresses itself in a great variety of options for incorporating biotechnology into industrial strategy.

Sensitivity to acceptability and consumer response, however, has made major companies wary of involvement in biotechnologies such as pituitary hormones or genetic modification of animals.

Intermediates and primary processing

The main products in this sector are yeasts, starter cultures for dairy processing, enzymes and a wide range of food ingredients including flavours, aromas and colourings. Enzymes are seen to have a double substitution potential: they may replace conventional chemical extraction processes or convert basic raw materials into substitutes for agricultural products.

Final foods

Two questions dominate innovation strategies in this sector: the likely consumer responses, and the potential of biotechnology to increase food quality. The industry faces two strategic issues: it must decide on the type and degree of its backward integration within the agrofood chain, that is, on its alliances with input and agricultural raw material suppliers, and it must also decide on its in-house, as well as its co-operative commitment to biotechnology R&D. Consumer response, and not technology push will shape industrial innovation, particularly in the field of genetically modified products.

c) Regions

The OECD area

The globalisation of competitive strategies is the main characteristic of leading agrofood companies. This means that national differences in the "innovation climate" with regard to agrofood biotechnology are becoming less important compared to individual in-house competence and various forms of R&D co-operation.

Differences continue to distinguish Japan's agrofood sector, which benefits from a more open consumer attitude and from a more comprehensive exploitation of the potential of fermentation technology, which results in useful synergies between foods and pharmaceuticals.

Developing countries

Recently, concerns have been expressed regarding the potential of agrofood biotechnologies to substitute for tropical crops and to stimulate competition between various raw material supply bases.

Current assessments of biotechnology impacts on developing countries are less alarmist. The increasing globalisation of the industry reduces local or national pressures for substitution, which, in any event, are less dominated by cost than by nutrition and health arguments. The latter do not, as such, militate against Third World crops. Moreover, increased demand for natural additives might partly make up for substitution losses by the Third World.

Urbanised developing countries are seen as the main long-term markets for agrofood biotechnologies. There are, however, institutional and legal consequences of multinational industry control of agrofood biotechnology. For example, companies demand better intellectual property protection than is presently offered in many developing countries.

d) Policy implications

To reduce the level of uncertainty, agrofood companies would like to see high priority given to more international harmonisation of safety principles and intellectual property rights.

R&D policies should stimulate new patterns of co-operation between the economic actors at different levels of the agrofood chain, focusing on quality improvements.

In order to exploit fully the flexibility of biotechnology to tailor products to specific markets and demands, consumer interests should be better integrated into production.

This requires a participatory view of consumer interests, rather than a merely reactive attitude.

Technology transfer procedures and institutions must take the dominant role of the large private corporations into account.

10. Public Acceptance and Intellectual Property

Among the constraints on industry and governments, it is public acceptance and patent protection which, along with safety assurance, are most often cited as delaying the development and diffusion of agrofood biotechnologies. The two obviously belong to different political and ethical categories.

a) Public acceptance

Public attitudes towards agrofood biotechnology and the exact reasons for consumer concerns are difficult to analyse despite a growing number of published opinion surveys.

As past surveys did not always distinguish between agrofood and other applications of biotechnology, the public often confused biotechnology with genetic engineering or with human reproduction technologies, or it associated it with unrelated food safety or public health concerns.

There are significant differences in public acceptance of biotechnology according to country, sex, educational level, religious practice, etc., and quantitative surveys have sometimes given contradictory results.

One of the most extensive surveys, comparing the attitudes in the 12 countries of the EC, was carried out in 1991 ("Eurobarometer"). It shows that biotechnology awareness is clearly correlated to educational level, personal income and GNP and, more importantly, that favourable attitudes to and objective knowledge about biotechnology are also clearly and positively correlated.

However, if the most highly educated groups show the highest acceptance of biotechnology, it must also be noted that the most active political opponents of agrofood biotechnologies belong to the same groups. Much of the opposition against agrofood biotechnology has grown as part of wider environmentalist concerns, which are mainly to be found among highly educated people.

It follows that governments and industry must better inform the public about new agrofood biotechnologies. Also, the participation of experts and people with no vested interest is required to assess the objectivity of this information. Public confidence in regulatory effectiveness needs to be significantly improved.

b) Intellectual property

Insufficient patent protection for biotechnological inventions in the agrofood sector has remained a prominent industry concern. However, the last few years have seen changes in the law or in its application, which have improved the patentability of products and processes in the agricultural and food sectors.

The main problem areas lie in the exclusion from patentability of certain inventions, which leads to legal diversities among OECD countries that are not helpful to invention and diffusion.

In Europe, legal protection for genetically modified plants and animals is not as straightforward as in the United States and Japan, in spite of recent efforts by the EC to raise protection to the levels of those two countries. European law, based on the European Patent Convention, prohibits the patenting of plant and of animal varieties in principle, in contrast to the United States which allows patents for both, provided patentability criteria are satisfied.

European plant breeders now accept the need to strengthen legal protection by various means, and after some original opposition, can see some advantages in patent protection for genetically modified plants. The European legal position with regard to animal patents is less clear, however. Opposition to patenting in these areas no longer comes mainly from legal circles, but from animal rights and environmental movements.

11. Economic Impacts

At the present stage in the development of agrofood biotechnologies, an attempt to assess their prospective economic impacts might appear hazardous. Nevertheless, some quantitative forecasts of biotechnology-induced yield increases have been published, and other likely or possible longer-term consequences can be evaluated in qualitative terms. Also, a case study of economic impacts is possible for one biotechnology product, bovine somatotropin (bST), for which quantitative economic forecasts have been made in a few countries. However, current assessments are still limited and fragmentary and need to be substantially improved.

a)   Uncertainties and methodological shortcomings

Government policies and general economic trends, neither of them easily predictable, affect the interaction between biotechnology and agriculture in major ways. At the present stage of the GATT negotiations (May 1991), future agricultural subsidy and price policies are particularly difficult to anticipate. A reduction of agricultural protectionism and high prices will affect biotechnological innovation and diffusion. Price reductions in high-price countries or sectors (e.g. milk, sugar) could slow down biotechnological innovation and diffusion during a period of structural change and transition. The ensuing greater efficiency, however, could lead to faster adoption of innovation later on. Also, by reducing agricultural inputs and/or costs, biotechnology could encourage the structural adjustment of agriculture.

There is no single biotechnology, but a wide range of technologies, varying in technical feasibility, profitability, etc., and thus with greatly varying economic impact potentials.

Some agrofood biotechnologies face particular problems related to consumer acceptance, environmental and safety concerns and intellectual property protection. Economic impact assessments have not given sufficient attention to the economic internalisation of such external constraints.

Biotechnology-induced quality improvements, the importance of which has already been mentioned. are not assessed in aggregate economic analysis, and thus are not integrated into impact assessments.

24

b) The wider context of agrofood biotechnology

Another limitation of impact assessments is that they do not go beyond farming, thus leaving aside the upstream and downstream sectors of the agrofood chain. These, however, have greater and steadily increasing economic weight compared to agriculture. As biotechnologies affect the entire agrofood chain, negative impacts on one part of the chain might be made up by positive impacts on another.

Links between biotechnology and the long-term, global food supply have not been clearly assessed. On the whole, the available general projections of global food supply are relatively optimistic about continued productivity increases with current "best practice" technologies, at least in developing countries. However, a wider spread of "best practice" technologies would require extensive change and diffusion of more traditional technologies, as well as institutional and policy adjustments on a global scale. Biotechnology offers additional options to reach the goal of an adequate global food supply at lower costs, including environmental costs.

c) Production

Aggregate estimates of impacts on agricultural production and market balances published for the USA and Europe reveal that biotechnology has a big potential to increase productivity, by decreasing costs per unit of output (e.g. less chemicals) or by increasing yields per unit of input. However, in highly developed countries, these biotechnology effects are likely to be partly offset by an expected decline of productivity increases due, among other things, to a slackening of more traditional technological factors. In some sectors, this decline can already be observed.

Specific impact assessments for bST show that even for a product with a technically big productivity-enhancing capacity that is relevant to a large sector of agriculture, the main direct result (in a quota system) will be cost reduction rather than production increases. If there are production increases, they will take time and, on an aggregate level, will be smaller than first expected. Current forecasts mention possible production increases of 10 to 16 per cent which would not arrive suddenly, but within a time frame of five to ten years. Thus, earlier forecasts of milk increases of 25 to 40 per cent, which have led to concern about further milk surpluses, may have been true for individual cows during trials, but were, as general economic assessments, widely exaggerated.

Over a longer period, however, bST could have a significant influence on milk production and farm structure if the new product is profitable, which is not yet the case in all circumstances.

d) Prices

Biotechnology, like other technical changes, will lead to downward pressure on food prices and thus to structural changes at the farm and employment levels. New markets, either for quality food products or for non-food products from agriculture, could alleviate this downward pressure.

While biotechnology could play a leading role in the development of non-food products from agriculture, conversion towards such products is hampered by policies supporting farm-gate prices for food, and by the fact that the price of non-renewable resources does not sufficiently internalise long-term scarcities.

e) Structure of the agrofood sector

Biotechnology will accelerate the ongoing structural changes of the agrofood sector towards increased vertical integration of agriculture with upstream and downstream sectors. Biotechnology adoption rates are expected to increase with management quality, which means that they are likely to increase with farm size (measured by income potential, not acreage). In general, biotechnology will tend to facilitate concentration in the upstream and downstream sectors of agriculture, and in agriculture itself.

For the time being, it is not possible to estimate the consequences of this concentration process on agricultural input prices. It may be necessary to monitor this process and to strengthen the bargaining position of agriculture with regard to the upstream and downstream sectors.

f) Trade and competitiveness

International trade and competitive advantages might shift in the long run towards countries which have both good capabilities in biotechnology and viable farms, if trade policies allow for such shifts. Economic impacts could be substantial. Experience with high fructose corn syrup (HFCS) in the United States has demonstrated in the past that the speed of adjustment to new biotechnology can be very high if incentives are adequate, although the substitution of sugar by HFCS has been strongly accelerated by sugar price protection in industrialised countries.

Shifts in trade and competitiveness must be monitored, as they can have both positive and negative consequences for developing countries. The danger of negative impacts on the Third World in general, expressed over the past few years, has probably been overestimated.

g) Employment

Biotechnology is expected to contribute to the continuing, long-term trends reducing the share of total employment in agriculture. International leaders in biotechnology might increase their international market shares and, thus, save employment in agriculture, if trade policies allow for it. However, this will come at the expense of the lagging countries which will lose more jobs, with no guarantee of a globally positive net effect. Employment reductions might be partly offset by new employment in downstream sectors, particularly the food-related services.

h) Agricultural policy

Biotechnologies are, in some way, related or relevant to nearly all major agricultural policy questions of our time. When they lie at the very heart of a problem, as is often the case, they may provide new and better solutions (reduction of chemicals, of factor inputs, of food contamination, more food for the Third World) or they may tend to aggravate it (market balances, loss of employment, more concentration), or they may do both at the same time.

As biotechnologies are likely to become more and more central to technical change in agriculture and food production, agricultural policies will have to cope with them and make use of them on a continuous basis. The main message is that biotechnology in agriculture is not a global issue, but one that must be addressed in a differentiated way, case by case.

12. **Policy Conclusions** (Formulated by the experts and a working group chaired by Sir Barry Cross).

1. **Biotechnology can be applied to all plants, animals and micro-organisms for food and non-food uses, thus affording the opportunity profoundly to improve the quality and efficiency of agricultural production.**

2. **New agricultural and food biotechnologies are extensions of traditional genetic modification practised in animal and plant breeding throughout civilisation, but further progress depends upon support of the basic sciences from which they have sprung.**

3. **During the next few decades, agricultural and food biotechnologies can make a vital contribution to global economic and social welfare for a rapidly expanding human population.**

4. **Biotechnology has the capacity to limit the damaging environmental consequences of some agricultural practices, deforestation and climatic change.**

5. **Contrary to earlier negative predictions, the diffusion of agricultural and food biotechnologies in the OECD area in the next ten years will be a gradual process without major, destabilising impacts on social structure or employment.**

6. **Public perceptions of biotechnology and safety issues remain potent sources of industrial uncertainty and must be addressed by improved communication among governments, scientists, industry, the media and the public.**

7. **Agrofood biotechnology could make a crucial contribution to the health and prosperity of the Third World, and it is therefore in the economic interest of both OECD and developing countries to devise, in concert, policies on research, intellectual property, safety, etc., that will enable the latter to build up their biotechnology capability.**

BIOTECHNOLOGY, POLICY AND THE FUTURE

As the prospect of global nuclear war recedes, the problems of feeding a world population expected to double in 40 years must loom large in the minds of responsible governments. The prediction of Robert Malthus two centuries ago that world population would outgrow food supply was not realized because mechanisation and the use of fertilizers enormously increased farm productivity. Present systems of agriculture, though capable of generating regional surpluses in developed countries, still leave a quarter of the world's people hungry and have little or no chance of filling the stomachs projected for the year 2030. Biotechnology in its various guises offers perhaps the best hope of avoiding a catastrophic impoverishment of mankind and our natural environment.

In a sense biotechnology has a history as long as breadmaking and brewing, but it received a strong impetus in the 1950s when the nature and function of the nucleic acids (DNA and RNA) were elucidated, leading to the unravelling of the genetic code and to recombinant DNA technology. It then became possible in the 1970s and 1980s to mass-produce genes in bacteria and transfer fabricated gene constructs to other organisms, not only bacteria and yeasts but also plants and animals, to produce valuable characters not accessible through conventional breeding. These revolutionary advances have raised bioscience to new commanding heights of biological control and given confidence that means can be found to dispel many intractable problems of life on earth – sterility, malnutrition, pests and diseases, energy shortage and environmental degradation.

In this volume the term biotechnology will have a meaning similar to the 1982 OECD definition: the application of biological organisms, systems and processes based on scientific and engineering principles, to the production of goods and services for the benefit of man (*Biotechnology – International Trends and Perspectives*, p. 21). It is thus wider than gene technology, the production and transfer of genes between organisms, and includes manipulations of cells and embryos, as for example in the production of monoclonal antibodies. It is important to realize that biotechnology is not a separate science, but a mix of disciplines – genetics, molecular biology, biochemistry, embryology and cell biology – transmuted into productive processes by coupling with practical disciplines like chemical engineering, information technology and robotics.

Biotechnology is ideologically neutral. It can be used to promote contrasting aims: to raise agricultural outputs or reduce inputs; to make luxury products or basic necessities; to replace chemical herbicides and insecticides or target them more efficiently; to upgrade pedigree flocks and herds or expand indigenous stock in undeveloped countries; to convert grain into biodegradable plastics or into methanol for fuel; to hasten maturity in livestock or prevent sexual maturation in locusts or in farmed salmon; to produce more nutritious and better flavoured foods or diagnostic tests for bacterial contamination; to engineer crops for fertile temperate zones or for semi-arid regions; to fight exotic virus diseases or build up populations of endangered species; to reduce production of "green-house" gases or utilise them in food production; to clone meat animals for particular markets or form embryo banks to maintain genetic diversity. Later sections give many examples of the extraordinary versatility of biotechnology.

There is of course a distinction between what may be technically feasible and what is actually achieved. An ever expanding knowledge base is necessary to get the best out of biotechnology. So far only very few useful plant or animal genes have been identified and much more needs to be learned about the complex system of molecular switches that control their expression. Many bright ideas, like nitrogen-fixing crops, may founder for lack of sufficient basic information. One of the most sought-after breakthroughs is an efficient method of genetically manipulating more of the major food crops such as wheat, beans, cassava, maize, and rice. There is no doubt that the remaining scientific problems will eventually be solved, but sooner rather than later only if help is given for research in the supporting sciences.

Which products and services biotechnology delivers in the next 20 years will depend upon the impetus of new scientific discovery and on the priorities in problem-solving perceived by the funding agencies as much as on the initiative of industrialists. The tensions between economic, political, ethical and social priorities are explored in later sections, but here it is worth making two simple points. First, virtually all modern biotechnology has originated in the richer OECD countries, and second, it is the other countries, those in eastern Europe and particularly those in tropical and sub-tropical regions that have the most urgent problems of health and food supply. It would seem desirable to look for a form of partnership that is less one-sided in character than some previous development programmes. Worth remembering in this regard is that many potentially invaluable genes may exist only in plants and animals that do not occur naturally in the developed nations. Who can say how vital these may become, for example in the event of climatic change resulting from the "greenhouse" effect?

A heavy responsibility lies on us all to ensure that the great intellectual insights into the molecular machinery of the biosphere that have made possible biotechnology in all its forms do indeed help to solve the world's problems, not create worse ones. This will demand the best efforts of our scientists and technologists and great wisdom from our politicians. Above all, there is a pressing need for better public understanding of the nature and potential of biotechnology in the agricultural and food context. We hope what follows in this volume will contribute to that understanding.

Part One

SCIENTIFIC AND TECHNOLOGICAL OPTIONS

Chapter I

PLANTS

1. Introduction

Plants provide directly or indirectly almost all the food of man and animals. They also supply industrial raw materials, for example, timber, pulp, paper, fibres, rubber, products for the chemical industries such as cellulose, starch, sugars, oils and fats, energy in the form of fuel wood, starch and sugars which are sources of ethanol, methanol, etc., and numerous valuable drugs, fragrances and other fine chemicals. Plant growth also has a massive influence on our environment. Because of all these roles, plants are a vital commodity for continuation of life on the planet and a major source of economic prosperity for almost every country. For these reasons policymakers should be continually developing policies for the use of plants to protect the earth's environment, to feed the growing populations, to satisfy man's ever increasing non-food requirements and to develop safe and efficient industries. Biotechnology is playing and will play a much greater part in meeting these objectives.

The need to consider urgently and continuously strategies for improving plant production and quality has many dimensions:

i) as pressure mounts to feed an expanding world population there is the need to grow more and more food globally while causing less damage and pollution to the environment;

ii) world population growth is occurring at faster rates in places where, in general, food production is inadequate or is vulnerable to environmental, economic and political changes;

iii) predictions of climate change suggest that the continued production of long-established crops in some regions cannot be guaranteed if, for example, droughts and high temperatures become more frequent;

iv) surpluses in many OECD countries are not substantial when global needs are considered, and many OECD countries are planning to reduce total production and to use agricultural land for other purposes;

v) there is now almost universal backing for the concept of managing total land resources and the adoption of crop production systems that are sustainable on good and more marginal lands using optimal amounts of chemicals and acceptable inputs of energy;

vi) a stable supply of (fossil) fuels at acceptable prices is not available to many countries;

vii) new scientific opportunities are arising through biotechnology;

viii) the private sector in plant breeding in OECD countries has changed recently, as some of the multinational agrochemical companies have invested heavily;

ix) there is growing public awareness in OECD countries of factors affecting food safety;

x) with food crop surpluses in OECD countries, alternative industrial uses for crops can be sought.

An overall scheme for the exploitation of plants is illustrated in Figure 1. The farmer and forester are central because they provide the raw materials for the consumer, the food industry and the diverse array of non-food industries. They manage the crops and they also manage the land resources. These activities are a very important contribution to crop production, crop quality and the sustainability of an agricultural system.

Plant biotechnology will make major new impacts on the inputs into agriculture as well as on the products and how they are utilised by industry. It will also provide novel means for monitoring and managing better the effects of agriculture on the environment and human health. In the past, scientific and technical discoveries have contributed much to improving agriculture. Today's and tomorrow's biotechnology will continue this process. Biotechnology is particularly powerful because it includes measuring and changing biological molecules and genes. The properties of crops can now be changed more efficiently in defined ways to meet specific targets set by man to improve production,

Figure 1. **A scheme illustrating the production and use of plants**

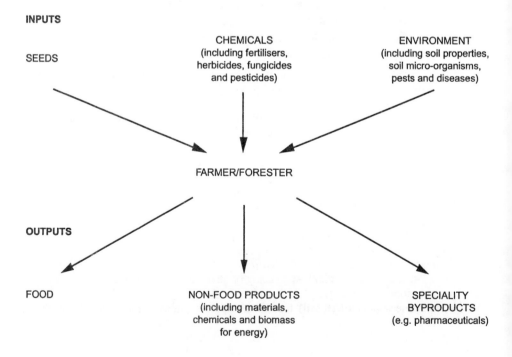

INPUTS

SEEDS

CHEMICALS
(including fertilisers,
herbicides, fungicides
and pesticides)

ENVIRONMENT
(including soil properties,
soil micro-organisms,
pests and diseases)

FARMER/FORESTER

OUTPUTS

FOOD

NON-FOOD PRODUCTS
(including materials,
chemicals and biomass
for energy)

SPECIALITY
BYPRODUCTS
(e.g. pharmaceuticals)

products, industrial process and the environment as well as to monitor disease, toxic molecules and the quality of plant products. Already plants and insects resistant to viruses have been created where no resistance previously existed, thus offering the opportunity to reduce the amounts of toxic insecticides used in agriculture. It is extremely fortunate that, against the pressing needs listed above for improving plant production and quality worldwide, a new portfolio of scientific opportunities has emerged to help solve the problems.

This chapter summarises the role of biotechnology, first in providing new ways of measuring crop quality, uniformity and contamination of produce with pathogenic organisms or chemicals, and second in finding ways of modifying crop quality to serve existing industries and consumers better and to stimulate new industrial uses of plants. These issues are especially relevant to OECD countries where, as noted above, there is both the opportunity and the need to give attention to food quality, safety, environmental effects and the industrial exploitation of high-value products. It also highlights biotechnology in plant breeding for better production on the farm via improved seeds and plants. These aspects are relevant to all countries and to all the points listed above. The role of biotechnology in converting harvested plant material into foods and other products is described in the separate chapters dealing with food and non-food products.

2. Diagnostics

In many OECD countries there is support for placing more emphasis on quality control of harvested crops in relation to food safety and suitability for industrial usage and also on maintaining the quality of the environment by the use of agricultural chemicals that are not detrimental to the ecosystem. As this concern is being translated into legislation, there is a need for defined standards and simple tests that can be used in field or factory to assay for quality and contamination.

a) For molecules and diseases of concern

Tests for ions such as nitrates, nitrites and heavy metals in soil and water are likely to remain chemically based although sensitive biotechnological tests are available. The routine use of such tests could for instance optimise fertilizer application and reduce applications where appropriate. The use of immunochemical technology based on antibodies for pesticide residue analysis has now advanced beyond the research stage and commercial lists, including single ELISA tests, are available to detect the herbicides atazine, simazine, propazine, cyanizine and thiobencarb, for example.

The development of immunoassays for analysis of pesticides, hazardous chemicals, industrial by-products and natural toxins in products and the environment is of great significance and illustrates a major potential benefit of biotechnology. Tests can be increasingly automated, as existing robots are suitable. It is to be expected that biosensors will be developed for environmental monitoring using specific antibodies coupled to electrical signal generators and easy-to-read output devices.

Tests on crops for contamination by disease organisms such as mycotoxins can now be performed rapidly by biotechnology-derived kits based upon monoclonal antibodies or nucleic acid probes that recognise sequences specific to the contaminating organism. Consequently, specific diseases can be detected in a crop at a very early stage and enable

Table 1. **Useful diagnostics from biotechnology**

	Benefits
Early detection of fungal, bacterial or viral plant pathogens	Optimal use of chemical control treatments
	Increased yields
	Improved crop quality
	Better product storage
	Safer foods
	Safer movement of germplasm between countries
Assay of plant type and quality for premium payments	On-farm and industry gate checks
	Stimulation of new uses of crops by industry based on specific qualities
	Checks on crop uniformity and legal ownership of germplasm

the farmer to choose the appropriate chemical and use it at low concentrations (Table 1), thereby reducing the amount of chemical sprayed into the environment. This technology can be helpful to the farmer and society, but needs to be developed into ''user friendly'' forms for the large number of agricultural diseases. The first such kits were marketed in the United States several years ago to detect destructive fungi in golf course grasses.

Tests to detect viral pathogens are also already in use in a few breeding companies and some quarantine laboratories. For example, the search for potato viruses X, Y and leaf roll in large populations of plants to find the virus resistant types, is being carried out routinely by advanced breeders using antibodies or nucleic acid probes to replace the more time-consuming and inefficient field-based observations of whether plants look infected. In the potato seed trade as well as in the companies selling stocks of many plants and bulbs, it is important to be able reliably to certify stocks as ''virus free'' because even cryptic viruses may limit yield. These same rapid tests are therefore extremely useful to define the phytosanitary quality of seed products for sale and their future yield potential. They are in routine use in the Netherlands, for example, to certify potatoes, bulbs, some vegetables, fruit trees and ornamentals. To move potatoes, as well as many other plants, across national borders, it is vital to know whether they are disease-free or not in order to control the spread of diseases. Methods based upon growing the plants in isolation houses for months to observe whether they are healthy are very inefficient. Rapid biotechnological methods are reliable, sensitive and make for safer, more efficient crop breeding and trade. The design and application of such phytosanitary tests needs to be stimulated by industry, public institutions and governments.

The value of harvested products is dependent not only upon freedom from diseases that may be harmful in food chains, but also on freedom from diseases that may stimulate rotting during storage. Rapid tests to assay for the fungi and bacteria responsible would enable appropriate treatment or early disposal, thereby increasing the quality and value of the produce for producer, industrial processor and consumer.

b)   For crop quality and industrial use

Use of crops as raw materials for industry is often based upon specific properties of the harvested parts. Rapid tests were developed years ago to assess, for example, sugar content in sugar beet, protein content and quality in breadwheat samples (see section 6) and nitrogen content in barley for brewing. Crops are sold on the basis of these assays. As research uncovers the role of specific proteins, genes and other molecules in industrial processing, simple tests to assay such molecules can be devised using biotechnology. Exploitation of such tests will aid the selection of suitable raw materials by industry and encourage moves towards the position where the relevant molecules are the commodity, not the plant material. For example, when plants are grown as sources of specific enzymes for industrial use, as in washing powders, then the amount and concentration of the enzyme, rather than the species or other attributes, are of critical importance.

The increasing emphasis on this diagnostic approach by the major industries is emphasized in Chapter V. It has significant implications for the research community, plant breeders (see section 6), agricultural produce users and consumers. The importance of developing valuable market sectors based on quality characteristics rather than plant species, is well illustrated by the EC wheat market in which animal feed wheats are in surplus but wheats with high gluten content suitable for bread-making are in deficit. Biotechnology tests can establish the uniformity of seed or plant tissue samples with respect to valuable properties and also provide ways of distinguishing a plant variety from all others (genetic fingerprinting). This ability to supply proof of identity provides an incontestable standard for plant variety rights and patenting defences.

In summary, there is a great opportunity for farmers, industries and societies world-wide to benefit from an ability to assay molecules and organisms with biotechnological methods. The exploitation of such assays will stimulate the efficient production of new and safer products, of higher quality. It would also permit legally defined standards for the protection of food quality and the environment.

3.   New Genes and Variants of Existing Genes for Crops

Plant improvement for better performance in the field or for subsequent utilisation by the food or non-food industries is achieved by the genetic modification of existing plants, followed by screening and selection of progeny to find those with preferable properties, than by evaluation of the performance of the new plant types in a range of appropriate environments. Biotechnology has much to offer to improve each of these steps in all breeding programmes, especially by providing valuable genes to solve problems and create new opportunities.

Since the beginnings of plant improvement by man, he has sought to widen the germplasm base. The history of plant breeding contains many notable examples, such as the "improvement" of maize to a crop very different from its wild ancestral forms by the former inhabitants of Central and South America and the selection of breadwheat containing chromosome sets from several different species. Such genetic modifications have proved decisive in feeding the world population to date. Recent biotechnology advances enabling us to purify single genes and insert them into crop plants look potentially to be another decisive breakthrough with wide application to many crops. Evaluation of modern plant biotechnology is assisted by consideration of the long history of modification of crop germplasm. The ever-widening sources of genes utilised are elaborated on below to reinforce this perspective.

a) *Sources of genes*

In conventional plant breeding, genes pre-existing within a species are brought together in new combinations by making sexual crosses (Figure 2 and Table 2,A). For the most part the genetic variation comes from mutations that occurred long ago in natural populations or during cultivation. Regardless of the genetic novelty introduced by the biotechnologist, most of the genetic variation required for crop improvement will remain what has earlier evolved in the species. This is why it is very important to keep large representative stocks of the natural genetic variation of key crop species. Such collections of germplasm are the source of individual gene variants and of adapted combinations of variant genes. Solving problems of plant production in novel environments will depend upon having such variation. It cannot be recreated once it is lost. More resources should therefore be put into germplasm collections maintained as seeds, as explants in cold storage or cryopreserved in liquid nitrogen. Biotechnology has led to new diagnostic methods to evaluate similarities and dissimilarities in germplasm. These methods therefore offer valuable help for the major problem of what plants to keep, what to discard and what constitutes a representative germplasm collection. They should therefore be introduced into all germplasm collection centres.

Valuable new genes can sometimes be introduced into crop species from a related species by making sexual crosses and nurturing the progeny embryos in special ways to enable them to survive, develop into plants and then be backcrossed to the crop species (Table 2,B). Progeny in which the foreign chromosome segment carrying the desirable genes has been introduced into a chromosome of the recipient crop species are then selected. For example, a gene conferring resistance to the fungal disease eyespot was introduced recently into commercial wheats on a chromosome segment from the wild relative species *aegilops ventricosa*. Also, surprisingly, maize pollen can now be used to fertilize wheat and barley egg cells, but the maize chromosomes are invariably eliminated from the resulting embryo so it is not yet known if this can be developed as a route for transferring maize genes to wheat plants across the species barrier.

New crop species have been created by crossing two species and maintaining the complete set of genes of both species in the progeny. This is how pasta and breadwheat species were created. More recently the new crop triticale was made by making fertile hybrids of tetraploid wheat crossed with rye. This crop, grown in many countries, has attributes of both parental species and grows better in poorer environments than wheat.

Foreign chromosome segments containing thousands of genes can be introduced by techniques which avoid the requirement for sexual compatibility, essential in the methods for introducing new genetic variation described above (Table 2,C). When single cells are created from plant material and their cell wall removed, it is possible to fuse the resulting protoplasts from dissimilar species and sometimes recover whole plants, using tissue culture methods, that contain chromosomes or parts of chromosomes of both species. Usually the resulting plants are of poor quality by agricultural criteria, but with more research there would be valuable commercial opportunities for gene transfer here. Of greater importance today is the transfer by these methods of genes not from the nucleus of the cell but from cytoplasmic organelles (mitochondria and chloroplasts) into another species. For example, transfer of organellar mitochondrial genes from male sterile radish

Figure 2. The breeding cycle showing conventional methods of gene transfer and evaluation and recently exploited molecular biological methods

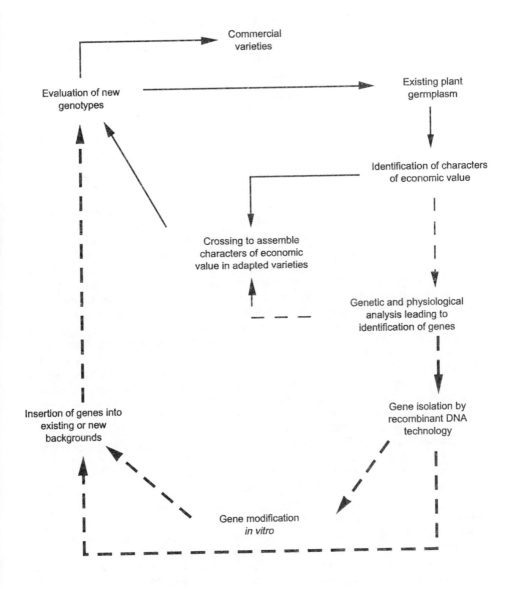

Long-established processes of genetic modification by sexual crosses.
Recently developed processes of genetic modification involving the transfer of single or a few genes.
Analytical research procedures that have been greatly stimulated by biotechnological techniques.

Table 2.   Sources of new genes for plant improvement

A   Other plants of same species. Germplasm banks are vital stores of such genes. Blocks of genes are transferred by sexual crossing.

B   Plants of related species, which are sexually compatible with the crop species. Blocks of genes are transferred by sexual crossing.

C   Unrelated species via fusion of single cells and regeneration of a whole plant from the fused cells. Blocks of genes are transferred including cytoplasmic genes.

D   From any species or synthesised in the laboratory; introduced into a single cell and the cell cultured into a whole plant. Single, specially designed genes are introduced.

to replace the equivalent genes in oil seed rape have created rape plants which are sterile because they do not produce pollen. To produce seed they must be fertilized by pollen from another plant. Thus the organellar replacement has created a commercially valuable way of making F1 hybrids, which frequently show yield advantages and offer companies a more attractive route to operating profitably.

Finally, new genes from any organism or synthesized in the laboratory can be introduced by the most recently developed techniques, in which single genes are introduced from the test tube into a single cell and the cell is cultured back to a whole plant (Table 2,D and Figure 3). Every cell in the progeny plant contains the new gene. The gene can be delivered directly into a protoplast or, in some species, can be transferred into cells by the natural delivery system evolved in a plant pathogen *agrobacterium*. This has been used to insert genes into a wide range of species; for example all the non-cereal species in Table 3. For those species in which it has proved difficult to regenerate whole plants from a single cell, such as the cereals wheat, barley and oats, the genes are being propelled on particles at high velocity into groups of cells, e.g. embryos, that can regenerate into whole plants. The gene to be inserted can be given regulatory information to ensure it is expressed only in particular parts of the plant. Its protein product can be modified, by modifying the gene structure, to give it optimum properties. These techniques offer a fine opportunity to modify species in desirable ways. However, commercial applications are at present severely limited by the number of valuable genes that have been identified and isolated. The international programme, just beginning, with the bold and ambitious aim to identify all the genes in the model plant, *arabidopsis thaliana* should greatly increase the number of gene functions available to modify crop plants.

Such research will also be aided by the more extensive and comprehensive identification of yeast and animal genes. Conservation of key functional components of crops between such diverse organisms enables functions to be recognised in unknown genes in other species and also to isolate the genes. This is facilitated by the world-wide availability of interactive, up-to-date computer databases containing the DNA sequences of all genes characterised. Such databases are a central aid in biotechnology today and influence considerably the rate at which known genes are available for insertion into plants.

Many plant improvements require loss or inactivation of a gene. Progress is beginning for some species in developing methods to target genes into chromosomal DNA with a related sequence, thereby inactivating the endogenous gene and achieving its replacement with a different version.

Figure 3. Scheme for modification of plants by insertion of a single gene to create insect resistant plants

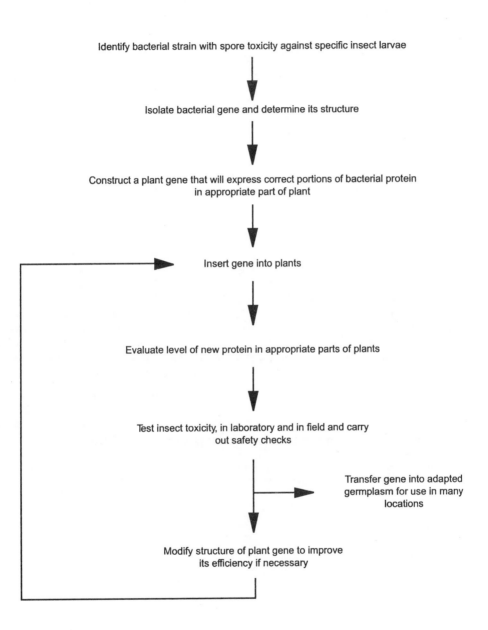

It is important that policymakers and the general public realise that genes are very rarely transferred intact from one species into another in these single gene transfer programmes. Usually a gene is redesigned in the test-tube to ensure appropriate activity in its new host. Thus the general formula "moving genes between species" associated with molecular genetic modification should more accurately be expressed as "adding redesigned genes". This distinction is important because the molecular biologist has deliberately created the specific attributes of the gene and therefore knows most of them very well. This contrasts with the more conventional approaches to genetic modification described above (Table 2 - A, B, C), where large numbers of unknown genes are transferred intact between species without public scrutiny or legislative control.

No plant was genetically modified by the insertion of a single gene using molecular techniques until 1982, and maize was added to the list only in 1989/90. The rapidly growing list (see Table 3) shows that research progress has been very rapid; several new species are being added to the list each year. The methodological difficulties lie with regenerating whole plants from the genetically modified cell, not with stable introduction of genes into the plant cells. More research is urgently needed to extend the number of crops that can be genetically modified in this way and to increase the number of private and public institutions with this capability. At present the application of biotechnology to plant improvement is severely limited by these deficiencies.

It is indisputable that the modification of existing germplasm by valuable genes or gene combinations from whatever source is essential to achieve the necessary adaptation of existing plants to meet future targets of the developed and developing countries. However, much research is needed to increase the efficiency of germplasm improvement to the necessary levels. For most crops in the world, especially those of the developing countries, relatively little multi-disciplinary breeding has been carried out, and so there is ample opportunity to improve plants using already well-established procedures. However, the knowledge that is gained on the species which are the subject of most research – maize, wheat, rice and model non-crop plant species (species of no commercial value

Table 3. **Crops into which single purified genes have been introduced**

The statement that a crop has been transformed by insertion of a gene only implies that one or a few laboratories have achieved the breakthrough. For most crops very few laboratories have the skills and, as yet, very few breeding programmes are using germplasm modified by such gene insertions. This approach to modifying germplasm is therefore in its infancy and still severely limited by technical difficulties and resources.

Cereals	Fibre	Legumes and Oil Seeds	Horticulture	Pasture
Rice	Cotton	Linseed	Carrot	Lucerne
Maize	Flax	Soybean	Cauliflower	
		Rape	Tomato	
		Sunflower	Tobacco	
			Sugarbeet	
			Potato	
			Lettuce	
			Cucumber	
			Celery	
			Strawberry	

but ideal for fundamental research) – will be useful to accelerate the efficiency of breeding programmes that have a weak history. Time is not on our side, and there is an urgent need to bring the cultivars of non-OECD countries up to the production standards of OECD countries in terms of yield and efficiency.

The problems to overcome in order to achieve this are many. Plant breeding and especially evaluation of crop performance must be carried out in the ecosystems where the crop is to be grown to satisfy human needs. Often the infrastructure to do this adequately is lacking in developing countries. In addition, supplies of desirable seed, fertilizers and other inputs are not available even where environmental conditions are more favourable for food production. The wealth to stimulate the private sector is absent, and often government financing is too low to ensure success. It seems inevitable that international resources, such as those cycled through the Consultative Group for International Agricultural Research (CGIAR) system, must continue to play a significant role in agricultural research for the developing countries and should therefore be protected. This need is even greater now that biotechnology research with considerable potential for the developing countries has arisen in developed countries.

b) Genetic modification: benefits and risks

The method of genetically modifying plants by the insertion of purified genes described above is controversial in some countries, and the release of such modified plants is becoming subject to legislation. There is concern that the foreign genes and their products might be harmful to human beings, to organisms in other food chains, to the plant species itself or to wild relatives to which it may pass the gene (see Chapter VI). Other people object because scientists are "interfering with nature", though as already mentioned the older, less well-defined ways of introducing foreign genes have not previously caused such alarm or demands for protective legislation.

There is no doubt that plants with properties deleterious to man and other food chains can be created by the insertion of single foreign genes; but this is also readily possible by long-established breeding methods in most species, and examples of such deleterious gene transfer are known. The possibility that genes, rare in a species but propagated in vast numbers in cultivated crop varieties, can transfer to wild relatives growing nearby has been with us since the beginning of agriculture. Also, the possibility that genetic variants of cultivated plants may become weeds, for example, by having longer-lived seeds is surely not novel to the products of biotechnology. There is therefore a strong argument for evaluating the particular products of plant breeding, the purposes for which they might be used, how they will be used and their possible effects in use on other organisms and environments, rather than focusing on how they were created, which has been the method adopted in many countries. In such evaluations the benefits should be assessed together with any adverse consequences before a decision to release or exploit is made.

It needs to be recognised that the risk:benefit conclusions may vary in different socio-economic and agronomic environments. Fortunately there is an enormous amount of experience in the world on the release and cultivation of plants with new combinations of properties. In most instances it is in fact much easier to predict the consequences of the changes resulting from the introduction of a single gene than of the hundreds or thousands of unknown genes that are introduced by the older procedures, which are not subject to the new legislation. Therefore, the risk:benefit analyses will be easier to carry out with plants modified by single genes. This appears not to have been appreciated by many policymakers, the media and the general public. There is a pressing need for greater public understanding of the principles of genetics, plant breeding, the history of agriculture and risk assessment to enable the release of new plants to be evaluated on a sound basis. Also, it should be remembered that the impact on the environment of any plant is related to how and on what scale it is used and not only to the properties of the individual plant.

Policymakers should keep in mind that agricultural and ensuing economic successes up to the present have come in part from the exploitation of specific, sometimes rare and foreign, genes, and this will continue to be so in the future. If plant breeding is held back by erroneously preventing the use of some sources of genes but not others, the consequences for solving the major food and agriculturally based problems of the planet like environmental pollution during the next century may be dire.

4. Improving Plant Breeding Programmes

The process of breeding is time-consuming and expensive, particularly because it is necessary to produce, grow and evaluate large numbers of plants for many generations. Where the crop has a long life cycle, for example with trees, then the evaluation programmes are especially long and expensive. Therefore, any procedures that help screen and evaluate germplasm more easily or earlier can make a substantial contribution. Indeed, they can make all the difference to a company decision on whether to develop the use of a particular gene or not. Biotechnology is providing many approaches to reducing time scales, and these are arousing considerable interest and investment.

a) Rapid identification of valuable genes

When a gene or gene product that must be maintained in a breeding programme is recognised, it can be assayed directly by protein or DNA molecular biology methods without the necessity of growing the plant in the field. If the gene has not been identified but a closely linked DNA sequence is available, then the latter can be used to hybridise to the homologous sequence on the chromosome in order to flag the desired gene in a breeding programme. For this, a molecular variation associated only with the preferred version of the gene must be found using restriction endonucleases which cut DNA at specific places and fractionation of the DNA pieces by electrophoresis. This detection of genetic variation is known as restriction fragment length polymorphism (RFLP) mapping.

In a more recent technical advance using a DNA polymerase chain reaction (PCR), short segments of chromosomal DNA are rapidly amplified from oligonucleotide primers, thereby providing easily visualised pieces of DNA, from specific loci, that frequently differ in length between genotypes. Programmes using these approaches are now well advanced to map hundreds of specific DNA sequences onto the chromosome maps of crop plants and to uncover linkage relationships between DNA sequences or combinations of sequences and valuable plant traits (see Table 3). As this work progresses, with possible elimination of the electrophoresis step, and the costs of application are reduced, then it can be expected that the efficiency of the progeny selection in breeding programmes will increase. It is desirable that the knowledge base for these new approaches be extended to crops in the developing countries.

New techniques in molecular genetics are leading to the recognition and isolation of important genes. These include inactivation of a gene by insertion into it of a foreign piece of DNA followed by isolation of the foreign DNA and with it the gene of interest. Such techniques are very powerful and will lead to many genes of great potential in plant modification. For example, genes regulating flower formation and dwarfism were isolated by this method in 1990.

b) New methods for hybrid seed production

In the developed countries and increasingly in developing countries, some crops are grown as F1 hybrids because of the better performance of such hybrids over inbred lines. Although theoretically hybrids are not essential to create such yield advantages, in practice the advantages are most easily realised in hybrids. Furthermore, commercial companies prefer to market hybrids because fresh seed must be bought each year and also because competing companies are prevented from exploiting directly germplasm that has been bred only at great cost. Hybrid seed, in maize and sugar beet for example, is created by growing together plants that cannot produce their own pollen with those that can. The seed harvested from those with sterile pollen must be hybrid. Several sorts of mutations can cause sterile pollen, some producing changes in the nucleus and others in the mitochondria. Prior to 1989, maize hybrids were created using the male sterility resulting from an aberrant mitochondrial gene in the "Texas" cytoplasm. This gene also conferred susceptibility to a fungal disease, so use of the mutation was discontinued after disease epidemics in 1969, and physical removal of the pollen-forming tassels was adopted instead, in spite of its cost, to create the necessary male sterile plants.

Where appropriate genes conferring male sterility are not available in a species they can be introduced by biotechnological methods. For example, mitochondria have been introduced from male sterile radish into oilseed rape by protoplast fusion to replace the rape mitochondria and have resulted in male sterile rape lines (see above). This approach is being extended to other species, especially the broccoli and cabbage vegetables related to rape. In other, more recent programmes, specially constructed genes encoding a ribonuclease have been inserted into one parent of the hybrid; they become active only during pollen formation and lead to infertility or death of the pollen, but this action can be overcome in the progeny plant by incorporating a gene encoding an inhibitor of ribonuclease to ensure hybrid seed production in the farmers' fields. Male sterile oilseed rape plants produced by this system are now in field trials. This biotechnological approach offers considerable hope for establishing hybrid seed production systems in

Table 4. **Species where detailed genetic maps to aid breeding have been created by biotechnological methods**

Cereals	Legumes and Oil Seeds	Horticulture
Wheat	Soybean	Tomato
Barley	Pea	Potato
Maize	Mungbean	Lettuce
Rice	Chick pea	Cauliflower
Sorghum	Rape	

species where none suitable now exist and for re-establishing a genetic system for hybrid seed production in maize. Breeders have tried unsuccessfully for many years, for example, to establish reliable hybrid seed production systems in wheat.

Hybrids in many species result because of self-incompatibility, the failure of a plant's own pollen to grow on its stigma. Insertion of self-incompatibility genes into species that lack them should also facilitate large-scale hybrid seed production, again provided that the system can be genetically arranged so that efficient seed production is not inhibited in the farmers' F1 plants. Research to achieve this is underway in *brassica* species now that the genes determining self incompatibility have been isolated.

The use of biotechnology to alter the breeding systems of plants can contribute very significantly to the production of high performance seed, and the structure of the seed industry can be radically altered if F1 hybrids are used where currently none are marketed, because companies will invest more in research, marketing and sales for a higher profit product that needs to be bought each year. In future, the limitations on whether or not seed of a crop is sold as an F1 hybrid will be the biology and genetics of the crop rather than the availability of a suitable male sterility system, because of the progress in biotechnology.

c) *Plant propagation and tissue culture*

Germplasm distribution, several steps in plant breeding and industrial uses of plant cells may be improved by the culture of parts of plants followed by the regeneration of whole plants from the parts. This area of biotechnology was established long ago and has been adopted widely but is now being expanded and applied to more species.

One major value of propagation of large numbers of plants from cuttings is that the genetic constitution of the plant is conserved. In many plants of hybrid origin the seeds are very heterogeneous and progeny plants would not have the characters specifically selected in the hybrid. Asexual propagation of the plants enables elite lines to be maintained and marketed. Tubers of potatoes, yams and stems of cassava and sugar cane are noteworthy examples of plants propagated by pieces. Propagation of meristems to produce a callus and then multiple embryos and plantlets in test tubes is now also commonplace for some species such as oil and date palm, strawberries and ornamentals. This enables many plants to be produced from a single meristem but is still relatively expensive. However, it has proved valuable for the routine elimination of viruses from cultivars of potatoes, sugarcane, cassava, banana and many ornamental plants and vegetables, for example, because for unknown reasons viruses do not readily replicate in fast growing meristematic cells. Maintenance of disease-free plantlets in culture has also proved to be a very useful way of storing elite germplasm, and transferring it across national boundaries.

Other more complex uses of tissue culture involve the culture of cells from the haploid microspores in the anthers to produce haploid plants and subsequently plants with the haploid genome doubled. This has been achieved for more than 100 species. Such plants are true-breeding and can be marketed directly (e.g. barley, rice, oil seed rape, potato). To achieve true-breeding lines without such a tissue culture step can take eight to ten generations and is therefore a much longer process.

In some species large numbers of cells can be sustained in suspension cultures and induced to form embryos and plantlets by manipulation of hormone levels (e.g. rice, carrot, potato). This is a promising method worthy of more research for proliferating species in a cost-effective way.

Cells in culture can be exposed to toxins, heavy metals, heat, cold and other stresses with some hope of selecting variant cells and subsequently plants that have improved resistances to such stresses. The process of culturing cells and plant pieces appears to increase the frequency of mutant offspring. This provides an opportunity to discover new valuable variants but creates a nuisance when the propagation of identical plants is desired. Valuable new genetic variants from tissue culture have been found in potato, tomato, pear and chicory as well as ornamentals. The viability of tissue segments including calli, somatic and zygotic embryos, anther and pollen embryos can be preserved by controlled freezing to very low temperatures after treatment of the tissue with cryoprotectants (eg glycerol and sugars or sugar alcohols). This enables germplasm of vegetatively propagated species or those with seeds of poor viability to be preserved in liquid nitrogen for long periods of time and successfully regrown after controlled warming of the tissues.

These biotechnology methods of tissue propagation and culture are being exploited substantially in many horticulture crops including ornamental flowers in developed and developing countries, but they need to be applied much more in forestry, where propagation of elite germplasm by such methods would greatly increase yield potential of forests. Application in these areas is discussed in Chapter IV.

5. Improving Plant Production

Crop performance in the field, defined as yield, is a very complex character and is affected strongly by environmental factors, soil type, external agents such as pests and

Table 5. **Plant characters frequently in need of improvement by exploitation of new genes in breeding programmes**

Increased drought tolerance

Increased salt tolerance

Increased cold tolerance

Increased heat tolerance

Increased disease resistance

Increased pest resistance

Herbicide tolerance

Increased nitrogen utilisation

Increased acid/alkali tolerance

Increased metal tolerance

Modified day length responses

Modified vernalisation responses

Increased photosynthesis/respiration efficiency

47

diseases, by the quality of agronomy and husbandry as well as by the properties of the plants themselves. Biotechnological methods can lead to increased yield by creating plants with attributes (see Table 5) that optimise exploitation of specific environments.

a) Herbicide tolerance

Early products of biotechnology in the category of genetically modified plants were herbicide-tolerant tobacco, potato, cotton, tomato, soybean and maize plants. Production of these plants with a view to commercial release has been criticised on the grounds that they will lead to more herbicides being sprayed into the environment, and some critics have condemned biotechnology in consequence.

Herbicides have proved to be very helpful in controlling weeds that compete with crop seedlings and significantly reduce yields. Added to soils before emergence of the crop, herbicides, if they are to be valuable, must be short-lived or not affect roots. Those added after emergence must not damage the crop but must kill the weeds. Some selective herbicides have been developed, but many crops are often not resistant to them. Therefore it would be advantageous if crops could be made genetically tolerant to a herbicide such as glyphosate, chlorsulphuron or phosphinothricin that eliminated all or most weeds. Biotechnology has led to the production of such plants by finding or creating a gene whose normal product, an enzyme in amino acid biosynthesis, is the target of the herbicide but which has become mutated so that its product is still functional but is now resistant to the herbicide. The gene, from bacteria, yeast or plant, has been modified for strong expression in plants and inserted into the crop plants. Field trials have confirmed useful levels of herbicide tolerance. In other cases, a gene whose product detoxifies the herbicide has been inserted into the crop plants.

Some herbicides are environmentally toxic and slow to degrade to harmless components. Such herbicides should not be used, and it would be commercially pointless to develop crops resistant to them. Other herbicides, such as glyphosate, are potent, short-lived and considered environmentally safe. The more extensive use of such a herbicide may have no undesirable side-effects by criteria accepted by most experts. Additionally, if crop plants resistant to a safe herbicide were widely grown, farmers would use it in preference to less desirable herbicides for which there is at present no safe substitute. It is difficult to predict the outcome of marketing crops resistant to "safe" herbicides, but it should lead to a reduction in the total use of undesirable herbicides as well as, of course, to improved agricultural efficiency.

Concern has also been expressed over the fact that where herbicides are used continuously, resistant weeds arise. This certainly occurs, but the resistant weeds can easily be eliminated by the use of a different herbicide. Also, in the absence of continued use of the herbicide, they do not usually have any selective advantage and so do not spread extensively in the weed population.

b) Plant morphology

Potential yield can be markedly influenced by the morphology of the plant. For example, the size and angle of the leaves to the sun can affect total photosynthetic efficiency. In small grain cereals substantial yield increases have been made by selecting plants in which a greater proportion of the accumulated biomass is transferred during plant senescence to the harvested seeds. Dwarfing genes, which reduce plant height, made the major contribution to the "green revolution" in wheat and rice because they reduced yield loss caused by the plants falling over, allowed more fertilizer to be applied without the plants growing too tall and enhanced other plant attributes such as the number of grains per ear. Identification of more genes which alter plant morphology will continue to be very worthwhile, especially in those crops that have received inadequate attention by modern breeders. Furthermore, it may be desirable to isolate or make different versions of known genes. Recent evidence suggests that the dwarfing genes used in Italian wheats, different from those used in Northern European wheats, are better for that climate because they allow earlier growth, enabling the plants to reach maturity before the heat and drought limit biomass accumulation.

Genes which affect flowering time or the growth response to day length and winter temperatures are known to influence the yield of many crops. Knowledge of the beneficial effects of these genes emphasizes the need to identify and isolate new variants of them so that they may be inserted into crops in which they are unknown. Genes of this type are currently being isolated from the model species *arabidopsis* in the hope that they will be useful in the search for variants effective in crop species.

Studies in crop genetics have revealed a host of other commercially valuable variant genes affecting plant morphology, for example, altered flower shape to increase cross-pollination efficiency, hairs on plant tissue to trap insect pests, and more prolific root structures to scavenge for water and nutrients. In trees, the number of shoots produced is clearly an important parameter, while in the cut flower industry the shape, colour and number of flowers are all critically important. Biotechnology can help to identify, manufacture and manipulate genes generating such valuable morphological features. For example, the par gene, which is activated by the hormone auxin, reduced apical dominance when overactivated in tobacco but caused dwarfism when active to only a low level.

c) Stresses

There is an urgent need to improve cultivars of many species to withstand adverse soils, stresses from heat, cold, toxic metals, salinity, and drought (see Table 5). All these factors are of considerable significance in limiting crop production worldwide. Perhaps one-third of all agricultural land is affected to some extent by salt deposits, and therefore genetic changes to increase salt tolerance are of great importance in many crops. Yet often they are unknown in the breeders' germplasm. The genetic control of plants' responses to such stressful conditions may involve many genes, and this makes it difficult for a breeder to bring together optimum combinations of all the genes into a single cultivar. Yield potential of a species is therefore never achieved. Furthermore, the genetic and physiological complexities obscure the biochemical basis of the limiting processes. This in turn produces problems in devising appropriate tests for evaluating genetically modified plants.

Nevertheless detailed genetic analysis of a wide germplasm pool often uncovers genes with major effects on such important characters affecting yield. For example, salt-tolerant strains of *elymus farctus* (a grass species) have been discovered that can withstand half strength sea water. Progeny from "artificial" crosses between these plants and bread wheat, possible only with special laboratory techniques, have produced plants with many attributes of wheat but which are also salt tolerant. Wheat plants containing the genes from *elymus* may, in the future, make an important contribution to agriculture in saline areas.

The biochemical and genetic basis of tolerance to heavy metals is known for some organisms, but it is not known whether these same mechanisms would apply to plants. There is a case for biotechnological approaches in plants based on fungal and bacterial models to achieve metal-resistant cultivars.

d) Nutrient utilisation

The yield of crop plants is very dependent on supplies of nitrogen. Legumes gain nitrogen by symbiotic associations with bacteria which fix atmospheric nitrogen, converting it to ammonia. The energy for this conversion comes from the plant. In some agricultural situations, legume yields can be maintained only by frequent applications of the appropriate bacterial strain. Considerable knowledge of the biochemical genetics of the nitrogen fixation process in these bacteria has been gained in recent years, but generally the genetic modifications have been done in laboratory strains which do not survive in the field. There is a need to apply these advanced technologies to agriculturally important strains. We also need to learn much more about how plants interact efficiently with nitrogen fixing bacteria. Selection and genetic modification of these processes may then lead to much more efficient nitrogen fixation and increased yields.

For species that do not interact symbiotically with nitrogen-fixing bacteria, and these include most of the world's crops, more efficient plant utilisation of artificial fertilizer is highly desirable to improve yields, while minimising costs and environmental damage. Knowledge of how plants scavenge nitrates and of the root structures most favourable for this should lead to new biotechnological solutions for plant breeders.

Micro-organisms, bacteria and fungi can provide useful compounds other than nitrogen in the soil to aid growth and combat disease. Few have been genetically improved and incorporated beneficially in a directed way into agricultural use to date. *Mycorrhizal fungi* that help phosphate accumulation are an example. Many micro-organisms that stimulate plant growth do so by suppressing the growth of other pathogenic micro-organisms. There are many opportunities for biotechnology here also. Some are noted below in section *f)*.

e) Disease

All crops are afflicted by damage from pests and diseases, and so serious are the losses in most crops worldwide (Table 6), despite the widespread use of chemical control agents, that improvements in disease resistance are one of the principal breeding objectives. Pests and diseases cause yield losses both in the field and in storage. Damage to the harvested products also reduces their value to the food processor and their appeal to the consumer. All plants are resistant to most pathogens and pests but each crop has a particular spectrum of pathogens and insects to which it is susceptible. The importance of disease control has led to widespread use of insecticides, fungicides and pesticides on crops at levels that are damaging to the environment. It is highly desirable to continue to strive to control pests and diseases genetically. Biotechnology has already made dramatic progress.

The phenomenon whereby the presence of a relatively harmless strain of a virus provides resistance to a subsequent attack by a more virulent strain ("cross protection") has been known for a long time. This gave rise to the idea of converting viral genes into plant genes to provide permanent "cross protection" to the plant species. In some of the first experiments the gene for the coat protein of tobacco mosaic virus was converted into

Table 6. Global estimates of losses due to plant diseases, 1987

Crop	Production (million tons)		Loss in production		Corrected value of loss
	1965[1]	1987[2]	Per cent[1]	Million tons[3]	(US$ billion)
Wheat	266	517	9.1	64.7	8.6
Rice	232	454	8.9	77.1	12.5
Maize	219	457	9.4	68.2	6.7
Other cereals	245	359	8.6	43.8	5.0
Potatoes	271	285	21.8	93.4	7.2
Sugar beet/cane	695	1 217	16.5	406.7	8.1
Vegetables	202	421	10.1	64.8	14.2
Fruit/citrus/grapes	142	326	16.4	74.8	15.2
Stimulants	10	11	14.9	2.9	3.7
Oil crops	95	134	10.2	19.0	4.5
Fiber crops/rubber	18	27	11.8	4.7	4.5
Total					90.2

1. Production and loss estimates used by Cramer in 1967.
2. *1987 FAO Production Year Book.*
3. Figures are estimates of losses as a proportion of potential yield in the absence of other constraints such as pests and weeds.
Source: James W.C., P.S. Teng and F.W. Nutter (1991), "Estimated losses of crops from plant pathogens", *CRC Handbook of Pest Management,* Ed. Pimentel, D.CRC Press, Boca Raton, FL, 2, 15.

an active plant gene and inserted into tobacco plants. Remarkably, plants with viral coat protein in all their leaf cells showed resistance to infection by tobacco mosaic viruses. The mechanisms are not fully understood, but the cell coat protein appears to prevent the coat protein being removed from the incoming virus, thereby preventing its proliferation to form an epidemic. Insertion of the coat protein gene into plants has been shown to confer some resistance to another ten viruses including potato virus, X, Y and leaf roll.

Another form of viral resistance has been created by insertion into tobacco of a gene expressing a copy of satellite RNA of cucumber mosaic virus. Satellite RNA is so called because it is an additional RNA not found in all strains of the virus, and its presence is associated with reduced virulence. Its presence in plant cells before subsequent infection with a virulent strain of the virus results in a considerable reduction in the severity of the infection. It is interesting to note that in China large acreages of peppers are treated with a mild satellite-containing strain of cucumber mosaic virus to protect them against subsequent infection by indigenous virulent strains. Thus the genetic modification solution simplifies and enhances a current agricultural practice.

Molecular biological knowledge is increasing rapidly on the mode of proliferation of viruses in cells, how they pass from cell to cell to infect large areas of a plant and how they are transmitted from plant to plant by insect or other vectors. This knowledge is enabling new genes to be designed for insertion into plants that should interfere with viral replication and transfer between cells or through insect vectors. Viral diseases are especially problematic in vegetatively propagated crops such as potatoes, yams, cassava, sugar cane and many bulbs where viruses are carried from generation to generation, without being eliminated during seed production. Other major viral diseases affect maize in Africa, and rice (rice tungro) in southeast Asia.

Plants resistant to insects have also been created by the exploitation of novel genes. The bacterium *bacillus thuringiensis* produces a protein in its spore which, when cleaved into its active form, is toxic to the larvae of insects. Different bacterial proteins have different toxicity spectra. Plant genes have been designed incorporating the bacterial coding sequence, inserted into tobacco, potato, cotton and tomato plants and shown to control effectively larval proliferation. The practical scheme adopted to create such plants is given in Figure 3 as an example of genetic modification projects by single gene insertion. Similar plants utilising a gene containing the coding sequence of a trypsin inhibitor of larval gut, isolated from cowpeas, and a lectin from peas have also been produced and evaluated. Such plants show there will be major advances in the control of pests and diseases emerging from biotechnology. They will be especially valuable not only because of the enhanced yield and quality but also because they should lead to big reductions in insecticide usage which is expensive and environmentally damaging. It is noteworthy that sprayed insecticides are not very effective at killing larvae in plant organs.

Other novel methods from molecular biology offer much additional promise in the medium to long term. In some instances plant genes that are activated in response to fungal or bacterial attack as part of the defence system of the plant have been defined. Such genes encode enzymes that can attack the pathogen, such as a chitinase which attacks fungal cell walls, or produce metabolites lethal to the invaded plant cell or cause lignification in order to prevent the pathogen invading neighbouring cells. These genes are activated by some signal produced in response to the attack by the pathogen. If we can discover the mechanism of this signalling system we should greatly enhance our ability to design new disease resistance genes. Such genes are likely to be specific for each pathogen, and since many resistance genes against a single fungal species are known within a plant species, there are likely to be several ways of genetically producing a plant resistant to a specific pathogen or pest. If several disease-resistance mechanisms could be built into a plant simultaneously, it would presumably be very difficult for a pathogen or pest to mutate to overcome them all. The crop plant might then become completely resistant to a fungal, bacterial or insect species.

To achieve these goals, which do now seem attainable in principle, it will be necessary to identify genes capable of being used to provide resistance in each crop against each pathogen, incorporating them into cells of the relevant plant species, regenerating plants from the cells and transferring the genes into the appropriately adapted cultivars for each location in the world. Obviously, results will emerge step by step, probably for the major crops first. Results on model systems may stimulate more directed attacks on the problems in subsequent research. This is illustrated by recent progress in genetic engineering of resistance against viruses and insects.

Plant breeders and biotechnologists must seek genetic variants that have lasting resistance to each of the pathogens and pests that attack the crop, but it is unfortunately common for these to mutate to overcome plant resistance mechanisms. These mutant forms then multiply selectively and reinstate epidemics. This problem is intensified by modern agricultural practices in which large areas are cultivated with a single crop, often of only a few cultivars over many years and when the population of natural predators is reduced by the use of chemicals or destruction of habitats. The ability of new races of some pathogens and pests to arise in response to new host plant resistance genes means that breeding for resistance is often a continuous process and one resistance gene has only a limited life in agriculture, unless new genes giving "permanent resistance" can be found. The stability of the new gene products of biotechnology is as yet unknown. It is therefore important that research continue to discover new disease resistance genes from any source that can be exploited in our principal crops.

f) Biopesticides

It has been recognised for a long time that pest and pathogen control exists widely in nature, and before the introduction of chemical pesticides, fungicides, etc., it was, together with husbandry procedures and host resistance mechanisms, a major component of disease control. Some of the natural pest and pathogen controls are due to pests and pathogens being attacked by other organisms (or chemicals produced within them) that are not directly harmful to crop plants. Such organisms therefore are and can be used as biopesticides. Some are listed in Table 7. At present, very few are in extensive artificially augmented use even though they are biodegradable and safe. This is so for a variety of reasons: many grow slowly on complex media and so are not easily produced in the large quantities that are necessary, many are fragile with a short life span, so are often not very

Table 7. **Biological agents with pest control activities**

Agent	Pathogen/Pest Controlled
Bacteria	
Bacillus popillae	Japanese beetle grubs
Bacillus thuringiensis – var kurstaki	Lepidoteran insect larvae
Bacillus thuringiensis – var israelensis	Dipteran insect larvae
Pseudomonas fluorescens	Plant fungal pathogens
Bacillus sphaericus	Mosquito larvae
Pasteuria penetrans	Plant parasitic nematodes
Viruses	
Insect baculoviruses	
Heliothis NPV's	Budworms and bollworms
Lymantria dispar NPV	Gypsy moth larvae
Neodiprion sertifier NPV	Pine sawfly larvae
Laspevresia pomonella NPV	Codling moth larvae on fruit trees
Fungi	
Colletotrichum glocosporiodes	Northern jointvetch
Alternaria casei	Sicklepod
Phytophthora palmivora	Mildweed vine
Metahizium anisopliae	Crickets, spittlebugs, stalk borers, etc.
Beauvaria bassiana	Wide variety of insects
Trichoderma spp	Rot fungi, soil-borne fungal pathogens, tree diseases
Neozygites spp	Spider mites
Protozoans	Grasshoppers and locusts
Nosema spp	.
Nematodes	Variety of insects
Neoplectana spp	
Heterohabditis spp	

Source: Carlton B. (1988), *Biotechnology and Crop Protection*, American Chemical Society, Washington, D.C.

potent and methods of application to maintain them on the crop are difficult. However, the application of biotechnology can be expected to overcome some of the constraints.

It is possible to discover the active products and genetically modify their production and structure to give the organisms much greater biopesticide activity. This is best illustrated by *bacillus thuringiensis,* spores of which have been used as a biopesticide for a long time. The recently acquired detailed knowledge of genetic variation affecting the structure and amount of the spore protein and its activity against different lepidopteran larvae has enabled many more potent strains of the bacterium to be created and made available for biopesticide use in agriculture. Also, the gene from *bacillus thuringiensis* that encodes the toxic spore protein has been inserted in pseudomonas strains that colonise the roots of crop plants such as maize, in an attempt to control larvae in the corn stem borer. It can be expected that with appropriate investment, the suitability as biopesticides of other species in the list of Table 7 could be substantially improved. This would enable biopesticides to play a more prominent role, together with safe chemicals, in an integrated pest and pathogen management system. Also, predators and parasites of crop insect pests, such as ladybird beetles, parasitic wasps and sterilised males could be made more effective by genetic modification.

6.   Improving Plant Quality

The quality of the harvested parts of plants affects their usefulness to man as food, or as a raw material for the non-food industry. Biotechnology will have a major impact on the quality and properties of harvested plant material and the way it is used. Over the past 20 years we have extended our understanding both of the needs and of the properties of the harvested parts necessary to supply these needs for the established industries. For example, plant breeders have selected for improvements in the brewing qualities of barley, in the breadmaking quality of wheat and for less toxic erucic acid levels in oil-seed rape for the food industry. Specific maize lines have also been selected for the food industry with modified starch/sugar levels that give altered degrees of sweetness. Potatoes have been selected that are particularly suitable for making potato crisps or boiling.

Tests to select for variants with better quality in breeding programmes are often similar to the industrial processes e.g. baking bread for breadmaking quality and micromalting of germinating barley seeds for brewing quality. In other cases such as oil quality, chemical analyses of seed constituents are used. The need to combine random genetic variation and time-consuming, expensive tests has limited progress, but there are clear commercial advantages to expediting work to comprehend the basis of quality attributes. For example, in wheat breeding it used to be necessary to convert seed into flour and bake a loaf to evaluate the suitability of the wheat variety for breadmaking. Then it was discovered that a flour extract made in detergent produced a sediment with a larger volume if the wheat variety was good for breadmaking. This enabled laboratory-based tests to be developed for small batches of seeds for breadmaking quality. Genetic and biochemical analysis of seed constituents subsequently showed that one kind of protein played a major part in determining the visco-elastic properties of dough and that electrophoresis of constituents from part of a single seed could reveal whether these proteins were of the right type or not. This discovery enabled breeders to screen single seeds to predict the breadmaking quality of the flour. Thus, knowledge of the precise constituent(s) of the harvested plant product responsible for the desired quality opens up more precise and rapid methods for predicting quality. The enormous scope for assessing plant quality by the development of biotechnological diagnostics was highlighted in section 2.

When a specific protein metabolic step or pathway is found to determine a particular quality, the possibility arises of interfering genetically to produce inherited modifications of that quality by gene insertion that are currently not available to conventional breeders. This is now stimulating research into the control of the starch, oil, protein and other biosynthetic pathways to discover how to create genetic changes that specify elevated levels of valuable end-products. By these methods it might be possible to produce, in oilseed rape for example, large quantities of the lauric acid oils that make coconut oil so useful in the food industry. This would provide a coconut substitute in temperate climates, though it could threaten the coconut oil exports of developing countries. The consequences of known mutations in the metabolic pathways for modifying plant starch and oil quality leave us in no doubt that biotechnology will provide plant tissues with a very wide array of new qualities in the future. Some of these changes will enhance the nutritional quality of crops as well as their suitability for new and old industries. Maize plants with elevated levels of the important amino acids lysine and methionine, which are limiting nutritional substances for monogastric animals, have already been produced.

In the horticultural industry quality is reduced by diseases that blemish foliage, fruit and flower blooms and reduce shelf-life of the product. Genetic control of diseases pre- and post-harvest is thus a major target to improve quality for the horticultural industry. In the cut flower industry, quality is also associated with novelty. The possibility of altering flower shape and colour by biotechnology is close at hand; the true blue rose is surely likely to be achieved by the modification of genes controlling enzymes in pigment biosynthetic pathways. Modification of petunia and tobacco flower colours has already been reported by the insertion of a gene from maize and also by other genes which interfere with normal pigment production.

Fruit and vegetable quality attributes are shelf-life, quality of taste, shape, general consumer appeal as well as suitability for the food industry. A more successful tomato might ripen in a controllable way, not go soft so rapidly, have a better flavour and in some cases be more suitable for soup or ketchup or other industrial uses. The rate of fruit ripening is determined by the ethylene produced in the fruit. Several companies are attempting to manipulate ethylene production genetically (the key biosynthetic steps are known) to reduce the rate of ethylene production to control ripening. Fruit softening is caused by the production of enzymes that break down pectin and other cell wall constituents. A target for genetic engineering is therefore interference with production of these enzymes; some success has already been reported. Other targets for the tomato are yields of solids and consistency when made into soups, as well as taste. Senescence and ripening is accelerated by ethylene in many vegetables and fruits, so the modification of ethylene production is desirable in any crops where shelf-life is critical, such as broccoli, which is required to stay green, and raspberries, which spoil readily during shipping. All of these objectives appear attainable with time, now that biotechnological approaches have become possible. Many other opportunities for the modification of quality by biotechnology are described in the food and non-food chapters (III and IV).

7. Constraints on Biotechnology Applications

a) Property rights

To invest in the production of a new plant cultivar there has to be a clear route to gaining a financial return for the company. Some OECD countries do this by giving cultivars plant variety (or plant breeders') rights which ensure return of a royalty to the breeder. This does not protect the variety being used by a competitor as a parent to produce other commercially valuable cultivars from which the first breeder gets no royalty. Many consider such schemes to be completely inadequate to provide suitable returns for the large investments necessary in plant biotechnology. It is now possible to patent individual genes in many OECD countries, and it is expected that royalties on each patented gene will accrue from any variety into which it is introduced. However, until the relationship between plant variety rights and patenting is clarified, together with the means of collecting the royalty, there may be inhibition of major investment by some companies. This is discussed more fully in Chapter VII.

b) Public perception

There are small but vocal sections of OECD societies that are arguing against the use of some processes in biotechnology to create new plants and micro-organisms. It remains to be seen how and when the public will buy genetically modified foods, and whether governments will insist on such products being labelled as "genetically modified". Also, will such products have to be evaluated as "safe" before they are allowed to be sold? If this happens, what will be the registration costs? All of these questions inhibit industry from committing larger investment into products and at the very least may hold up the flow of a return on money already invested. Chapter VI considers some of these problems in more detail.

c) Scientific limitations

Not many agricultural crops and trees are well-known genetically, with a wide array of well-characterised new genes available. There is therefore a serious shortage of valuable new genes for crop improvement. Also, most crops are difficult to transform with new genes from alien sources even in the best laboratories. Other crops where propagation by tissue culture is desirable do not respond in culture. Therefore, the application of biotechnology is limited by knowledge, human skills and resources. While it is possible to add genes to some species, it is only just becoming possible readily to replace genes in species; there is a need to be able to do this to modify some characters.

8. Prospects for Commercial Exploitation

Many of the recent biotechnology developments such as genetic mapping, diagnostics and propagation by tissue culture are already being exploited for commercial gain, and there will continue to be substantial growth and proliferation in the utilisation of these techniques.

Table 8. Recent biotechnological achievements for potato breeding

For Rapid Identification and Isolation of Valuable Genes:
1. A map of molecular markers covering the whole genome has been created.
2. Transposable elements from maize have been inserted and found to move.

For Diagnostic and Disease Control:
1. Nucleic acid probes and monoclonal antibodies have been produced to detect the three principle viruses X, Y and leaf roll, and nucleic acid probes are available to detect potato spindle tuber viroid.
2. The use of the diagnostics has been reliably incorporated into breeding programmes to improve efficiency.

For Hybrid Seed Production:
1. Genes regulating self-incompatibility have been isolated for future modification.

For Plant Propagation and Genetic Modification:
1. Haploids and dihaploids have been produced for the creation of diploids, triploids and new tetraploid plants.
2. Protoplasts have been regenerated into whole plants.
3. Protoplasts from different species have been fused and regenerated into whole plants.
4. Micro and macropropagation have been developed for proliferation of new plantlets.
5. Microtuber production has been developed in vitro for mass propagation of tubers.
6. New genes have been stably inserted into cells and transgenic plants regenerated.

Valuable New Germplasm Created:
1. Plants resistant to viruses X, Y and leaf roll.
2. Plants resistant to insects.
3. Plants resistant to fungal toxins.
4. Plants with altered starch levels and amylose/amylopectin ratios.

For the reasons given above as well as the relatively long time scales in plant breeding, multiplication and marketing, the time scale for financial returns is longer than in many other industries. Furthermore, when a new valuable gene is discovered or created, it is necessary to transfer it to many cultivars that are suitable for growing over wide geographical regions. These transfers, multiplication and marketing processes may be time-consuming and complicated to organise. In many OECD countries financial returns are influenced highly by government-controlled subsidies. These can be changed to affect the attraction of a crop to farmers and the stability of the market for long-term R&D.

The new methods in plant biotechnology that have emerged over the past decade are already utilised at a higher level of development in industry than in public sector laboratories. The multinationals and especially the agrochemical companies of the United States, Europe and Japan have taken the lead. The progress for the principal crops of potatoes, oil-seed rape, rice, maize and soybean has been dramatic, even though many problems remain to improve the efficiencies of the technique. Some of the achievements for potato are given in Table 8 as an example.

The first transgenic plants of cotton, potatoes, oil-seed rape, tomatoes, tobacco and soybean have been evaluated in small-scale field trials. Commercially valuable transgenic cultivars of these species exist with resistances to herbicides, specific viruses and insects.

Table 9. **Impact of plant biotechnology in developed and developing countries**

A. Current major developments associated with biotechnology	B. Likely consequence in developed countries	C. Consequence to agriculture in developing countries
1. Development and commercialisation worldwide dominated by multinational seed companies.	1. A major part of germplasm production controlled by a relatively few companies in the major crops.	1. Limits advances in major crops if access not granted to products of the multinational seed companies.
2. New ways of measuring quality of crops.	2. A diagnostic kit industry will be stimulated for farmers and industry to use; premiums may be established by such standards.	2. Unlikely to be adopted on a wide scale in near term.
3. New ways of measuring diseases and toxic contaminants of crops.	3 & 4. Farmers will use optimum levels of chemicals and be able to market products with known levels of "contaminants".	3 & 4. Unlikely to be adopted on a wide scale; chemicals often not readily available.
4. Means of optimising chemical inputs from measuring needs.		
5. Reduction in disease and pest control by chemicals due to use of new resistance genes.	5. Use of chemicals on farms will decrease. Productivity will increase.	5 & 6. If genes adopted, could be of major benefit to increase and sustain yields but use dependent on transfer from multinationals in spite of no profit return.
6. Modification of new genes to improve productivity and quality of crops.	6. Better varieties of plants will be available.	
7. New industrial uses of modified crop products.	7. Industry will use more plant products as raw materials; new industries will emerge.	7. Some examples likely if investment adequate and where crop sources are available.
8. More efficient plant breeding programmes.	8. Rates of variety improvement will be sustained.	8. Breeding programmes will be more efficient in major (externally supported) centres.
9. Patenting of genes and germplasm.	9. New systems for the return of royalties on genes and varieties to breeders will be established; seed will tend to be more expensive.	9. May be a hindrance to use of improved germplasm from developed countries.
10. More seed marketed as hybrid.	10. Seed will be more expensive but of better yield potential.	10 & 11. Of little benefit unless good local subsidised seed/plant distribution systems set up.
11. More propagation of elite pathogen-free germplasm by tissue culture.	11. Better seeds/plants will be available to more growers.	11. Better seeds/plants will be available to more growers.
12. Greater movement of disease-free germplasm around the world.	12. Breeding programmes will benefit.	12. Breeding programmes will benefit.
13. Desirability of production of transgenic crops is being debated by small sectors of societies.	13. Regulatory infrastructure will be established to monitor transgenic crops.	13. Regulatory infrastructures may not be set up; transgenic crops are likely to be used in germplasm; provides an acceptable improvement.

There has therefore been early success for these characters of the list given in Table 5. If there are no delays due to public concerns, registration, etc., then such cultivars could be on sale in quantity as early as 1995. However, cultivars suitable for a wide range of OECD countries would not be ready for wide-scale adoption until closer to 2000. It is possible that there will be some delays in marketing and further developing such products, so it seems likely that few transgenic cultivars will be in commercial use before 2000. Research to produce genes for many of the list of needs in Table 5 is unlikely to be successful for five to ten years. Therefore, it can be predicted that transgenic plants with many new traits will not make an extensive impact until after 2000. Between now and then the efficiency of the technology can be developed for more crops, in more institutions around the world, while attention is paid to the constraints addressed in the last section.

9.   Global Aspects of Plant Biotechnology

The preceding sections have described many agricultural activities where developments in plant biotechnology will have an impact. These are summarised in Table 9,A. Descriptions of the impacts are necessarily general in a report such as this; particular impacts will vary from crop to crop, and in different socio-economic groups and countries (see Chapter VIII). The exploitation of biotechnology will differ significantly among countries with widely different R&D capabilities and agricultural infrastructure. For simplicity, these are noted as "developed" and "developing" countries in Table 9, B and C, where the general impacts are compared.

As has been noted throughout this chapter, there are many benefits to plant breeding, plant production, the use of plants and care of the environment that can come from biotechnology, but, as with all technologies, detrimental consequences can also result. To avoid these, careful planning by industries and governments will be necessary. The high costs of R&D also suggest that applications of biotechnology to crops of poor profitability will not be done by the private sector and so must be done with government support. The less profitable crops are frequently those of lower sales volume, where the farmer saves the seed for use the following year or where his financial return is relatively low because the crop is not valuable to industry.

When the positive consequences of plant biotechnology for the developed countries listed in Table 9,B are assessed against the major concerns and targets for agriculture listed in the introduction, it can be seen that the efficiency of production should be increased, and the detrimental impact of agriculture on the environment should decrease. Overall, plant biotechnology will help increase the potential for plant and food production in an acceptable way in the developed countries. However, the extent will vary from crop to crop and from country to country and will be heavily influenced by many of the factors described elsewhere in this and other chapters.

The situation in the developing countries is, however, significantly different. These differences are very important, given that many of the major concerns of plant production relate to the developing countries. Survey of the points in Table 9,C shows that the benefits of biotechnology will not accrue in many developing countries unless valuable genes, germplasm and diagnostics become available to them, even though the direct financial returns to the originating industries of the developed countries might be minimal. It is in the interest of OECD countries to recognise the needs of developing countries and join with them to facilitate the transfer of technology and know-how to exploit plant biotechnology and its products. Without this, the agricultural wealth differences between developed and developing countries will widen further. The mechanisms to achieve technology transfer and training must involve governments, international agricultural agencies and industries working together. As noted in the table some genes, such as those conferring resistances to virus and other diseases, could have a major impact on yields in developing countries and help protect the environment if incorporated into the adapted germplasm. Their transfer should therefore be given highest priority; also needed in the developing countries are seed multiplication and distribution systems. Without these, new germplasm improvements will be of little consequence.

10. Further Reading

AUSTIN, R. B. (1985), *Molecular Biology and Plant Improvement,* Cambridge University Press.

COPPING, L. and P. ROGERS, eds. (1985), *Biotechnology and its Application to Agriculture,* BCPC Publications, Croydon, UK.

CUBITT, I.R. (1981), "The commercial application of biotechnology to plant breed", *Plant Breeding Abstracts,* Vol. 61, pp. 151-158.

GASSER, C.S. and R.T. FRALEY, (1989), "Genetically engineering plants for crop improvement", *Science,* Vol. 244, pp. 1293-1299.

GRIERSON, D., ed. (1991), *Plant Genetic Engineering,* Blackie, Glasgow and London.

HEDIN, P. A., J.J. MENN, and R.M. HOLLINGWORTH (1988), *Biotechnology for Crop Protection,* American Chemical Society, Washington DC, USA.

JONES, M.G.K. (1989), "Genetic manipulation of arable crop plants - present status and future prospects", *Agricultural Progress,* Vol. 64, pp. 1-18.

PERSLEY, G. (1990), *Beyond Mendel's Garden: Biotechnology in the Service of World Agriculture,* The World Bank, CAB International, Wallingford, Oxford, UK.

PERSLEY, G., ed. (1990), *Agricultural Biotechnology: Opportunities for International Development,* CAB International, Wallingford, Oxford, UK.

POTRYKUS, I., ed. (1990), "Gene transfer to plants: a critical assessment", Proceedings of the EMBO Workshop, *Physiologia Plantarum,* Vol. 79, pp. 123-220.

SCHUCH, W. (1991), "Improving fruit quality through biotechnology", *AgBiotech News and Information,* Vol. 3, pp. 249-252.

SILVER, S., ed. (1985), *Biotechnology, Potentials and Limitations,* Springer Verlag, Berlin.

VASIL, I.K. (1990), "The realities and challenges of plant biotechnology", *Bio/Technology,* Vol. 8, pp. 296-301.

WICKENS, G.E., N. HAQ, and P. DAY, eds. (1980), *New Crops for Food and Industry,* Chapman and Hall, London.

Chapter II

ANIMALS INCLUDING FISH

1. Introduction

Valuable traits in farm animals have been improved by selective breeding extending over hundreds of years. In the modern era the rate of livestock improvement has greatly benefited from advances in quantitative genetics and reproductive biology such as artificial insemination and embryo transfer. However, the annual genetic gain for desirable traits like growth rate, health, leanness or milk yield in ruminants is generally under 2 per cent, and since the generation interval in animals like cattle is very long, the results of technological improvements are only visible after a long period. Biotechnology has the potential to offer greater improvements, partly because modern techniques can help to shorten these intervals and partly because of the gain that can be achieved using the products of biotechnology (Table 10). With a world population of over 5 billion people and an estimated doubling during the next 40 years, countries of the OECD need to explore new ways to enhance food production and clearly have a leading role to play in this quest. The impact of new discoveries will not only be felt in developed economies, for half the world's 1.2 billion cattle equivalents reside in Africa and Asia, particularly in the Indian sub-continent.

Animal biotechnology is expected to have a significant influence on agriculture over the next five years, though it is well recognised that the time interval from basic research to application may be ten or twenty years. Advanced biological systems are being studied to improve:

Table 10. **Comparison of gains in milk yield per cow for different dairy technologies**

Technology	Annual gain in milk/cow[1] (kg)
Artificial insemination (AI)	50
AI plus sexed semen	55
AI plus embryo transfer	68
Bovine somatotropin	<1 000

1. Values represent average actual gains using artificial insemination and superior sires, which were approximately half the expected theoretical gain. Hence, the bovine somatotropin gain has been reduced by 50 per cent. It should be noted that the annual gain in milk/cow would be heritable with improved breeding (artificial insemination, sexed semen, embryo transfer). Annual gain from bovine somatotropin would occur only in the year of treatment (non-heritable).
Source: Bauman (1990).

i) animal quality, in terms of production performance or the controlled expression of introduced foreign genes;

ii) animal health, with respect to disease resistance; and

iii) animal welfare, by diagnosis of management stress effects.

The systems used to enhance the quality, health and welfare of domesticated farm animals include recombinant DNA (r-DNA) technology, monoclonal antibodies (MAB), cell fusion, and protein engineering. A number of commercial products have already reached the market, mostly in the field of diagnostic testing, and profiles of about 100 companies involved in veterinary biotechnology show that the industry is characterised by a large number of collaborative ventures. This emphasizes that research and development in biotechnology is often better done by pooling expertise than in isolation.

Animal cell biotechnology is an important interface between biological and medical science and will play an increasingly important role in the future development of veterinary and human medicine. Gene insertion for the study of foreign genes in new sites under the control of new switching mechanisms (referred to as promoters) is already yielding insights that are as relevant to human medicine and genetic disorders as to animal biology.

The ensuing sections summarise progress, opportunities, potential applications and commercial prospects arising from work with traditional farm livestock and with fish.

2. Animal Health

a) *Diagnosis*

There are various ways in which biotechnology can assist in the control of animal disease and the monitoring of animal health. The first is by the production of diagnostic agents such as monoclonal antibodies and DNA probes that can detect minute quantities of infective material in affected animals or endocrine and genetic abnormalities before clinical signs have become evident. Positive animals can then be segregated for further investigation. *In vitro* diagnostic tests can be used not only to detect the presence of the infective agent in body fluids or tissues, but also to monitor the host's response to the pathogen (by specific antibodies, blood components in milk, etc.), and to detect the occurrence of congenital disease. These tests offer an expanding field for fruitful investigation.

Rapid diagnosis requires highly sensitive detection systems utilising promising amplification techniques for vanishing weak signals derived from specific molecular interactions. Among companion animals, diagnostic tests are targeted for feline leukaemia, canine heartworm and rheumatoid arthritis; among poultry, large scale screening of avian reovirus, coccidiosis, salmonellosis and respiratory infections; among large domestic animals, trichinosis, mastitis, leukaemia, infectious bronchitis, brucellosis, babesia, transmissible gastroenteritis, rinderpest, Aujeszky's disease, blue tongue and trypanosomiasis. Collaboration of academic and government research laboratories with international agencies (e.g. Food and Agriculture Organisation, Rome; International Atomic Energy Authority, Vienna) has resulted in availability of diagnostic kits and computer software programmes that will provide access to epidemiological data bases for rapid analysis.

Opportunities exist for developing and improving diagnostic procedures for an even wider range of infections including foot-and-mouth disease, swine vesicular disease, African swine fever, classical swine fever, capri pox virus and African horse sickness. Nucleic acid hybridisation and sequence analysis are proving increasingly valuable not only in diagnosis (e.g. foot-and-mouth virus) but also for differentiating viruses in epizootics (e.g. blue tongue viruses). Though such techniques are in their infancy, potential applications have special importance for developing countries. Early diagnosis is an essential prerequisite for containment and elimination of epizootic viral diseases.

To grasp the speed at which benefits can be expected from the application of biotechnology, it is instructive to consider the growth of diagnostic procedures. Diagnostic products had one of the earliest impacts on animal health. By the end of 1989, over 120 test kits suitable for use by the veterinarian or farmer were available, the majority being "user friendly" with easy to read end-points. Fertility monitoring of large animals is currently the largest single area of application for diagnostic kits, involving measurement of fertility hormones in blood or milk (progesterone, oestrone sulphate, equine

Table 11. **Examples of diagnostic tests used to monitor fertility and launched between 1984-1989**

Company	Test	Species
Alfa-Laval Agri (UK)	Progesterone	Cow
Cambridge Veterinary Sciences (UK)	Progesterone	Cow (Dairy & Beef), horse, pig
	Oestrone sulphate	Pig
	Acute phase proteins	Dog
Hoechst Animal Health (UK)	Progesterone	Cow (Dairy)
Quidel (US)/ Clonatec (France)	Progesterone	Cow (Dairy)
Hoechst (Germany)	PMSG [1]	Horse
ImmunoCell (US)	Progesterone	Cow (Dairy)
IQ (Bio) (UK)	Progesterone	Cow (Dairy & Beef), horse, pig
IDEXX Corp (US)	Progesterone	Horse
Monoclonal Antibodies (US)	PMSG [1]	Horse
Noctech Ltd (Ireland)	Progesterone	Cow (Dairy & Beef), horse, pig
	Oestrone sulphate	Cow (Dairy), goat
Pitman-Moore (US)	Progesterone	Cow (Dairy)
	PMSG [1]	Horse
Probe-Tek (US)	Progesterone	Horse
RIA Ltd (UK)	Parturition	Horse
Sanofi Animal Health (France)	Progesterone	Cow (Dairy)
Synbiotics Corp (US)	Progesterone	Cow (Dairy)

1. PMSG: pregnant mares' serum gonadotrophin, now known as equine chorionic gonadotrophin.
Source: Bloomfield (1988).

chorionic gonadotrophin (see Table 11). Studies of sperm fertilizing ability with concomitant application of a variety of computerized tests based on examination of motility have identified means by which the quality of different batches of semen can be predicted.

There are big opportunities for technological innovations to increase the speed and improve the sensitivity and simplicity of diagnosis. Detection of oestrus in large animals still offers a major opportunity for new tests based on detection of minute quantities of hormones such as oestrogens or luteinising hormone in blood, milk or saliva. Tests for early pregnancy or the onset of parturition, in addition to oestrus, demand simple "on-farm kits" probably involving biosensor technology which can be used by farm staff. This technology has seen notable advances and promises significant improvements during the next decade. On-line monitoring in milking machines of hormones, electrolytes, antibiotics and cell numbers could provide opportunities for regular fertility monitoring and early detection of mastitis, but large scale application will depend on the solution of formidable problems involving membrane, electronic and physico-chemical technologies.

b) New vaccines

The second strategy is to use recombinant DNA technology to manufacture vaccines for prophylactic use. It is possible to make vaccines based upon an antigenic sub-unit of the infective microbe, which, though effective in stimulating the animal's immune system, is devoid of pathogenic influence. One of the first commercial sub-unit vaccines was that against hepatitis B virus in man, and there has been progress with an antirabies vaccine produced in a similar way. A further refinement is to produce vaccines from synthetic peptides duplicating fragments of one of the virus proteins, as has been achieved for the foot-and-mouth virus. When fused or hybridised with the hepatitis B core particle, the induced immunity to foot-and-mouth disease was still further enhanced.

Among the classical vaccines, those for clostridium, Marek's disease, cholera and coccidiosis comprise the largest single agricultural vaccine markets. Their efficacy frequently depends on active immunity whereby the animal's immune system responds with antibodies and cellular reactions primarily against surface components of the pathogen. The first veterinary vaccines using recombinant DNA technology were Nobi-vac LT, K88 and K99, designed to protect calves and pigs against enteric colibacillosis. The first genetically altered virus vaccine (Omnivac-PRV) was a live virus vaccine for the prevention of Aujeszky's disease in pigs (pseudorabies) and companion diagnostic tests have been launched to distinguish vaccinated from naturally infected animals. An example of a genetically engineered vaccine against a major disease in poultry which seems a real prospect for the future is the one against fowl typhoid (Table 12).

Novel sources of antigen contribute to the new interest in vaccine design, especially in relation to parasitic disease. Recent work has successfully used material from early developmental or infective stages or from especially vulnerable organs of the parasite such as the gut lining. A high degree of active protection against certain helminths has been produced by these means. Furthermore, immunogenic vectors (e.g. vaccinia, fowl pox virus, non-pathogenic salmonellas and eimerias) make "natural" immunisation with unnatural components possible. Recombinant fowl pox virus vectors have recently been developed to induce immune responses in the absence of virus replication. Thus, pathogen proteins can be grown in safe systems so the need for bulk pathogen work can be circumvented, while newer techniques of peptide synthesis are making peptide-based

	Description	Comment
Novel Step	Improved fowl typhoid vaccine	Avirulent mutant producing virulence proteins to stimulateimmunity
Commercial Consequence	Reduction of fowl typhoid in poultry production – especially breeding and laying flocks	
Current Market and Technology	Use of live attenuated vaccine with significant side effects	Main market is overseas; particularly Latin America, India and Middle East
Major Benefits	Reduced side effects: mainly lowering of egg output. Improved control of fowl typhoid, with extension beyond the present 12 weeks.	Ability to use genetically marked strain as vaccine.
Commercial Infrastructure	Well developed in animal health and poultry sectors.	
Sectors Affected	Animal health. Poultry breeding and production.	
Next Technical Steps	Isolate and clone genes. Assess clones as vaccines.	
R&D Base	UK public laboratory.	No other work known.
Time and Cost to Market	2-3 years and £200 000.	
Competing Technology	None.	
Lateral Uses of Technology	Similar systems for salmonellosis and other bacterial disease could be developed.	

Source: Prospect Management Services (1988), *Biotechnology for Businessmen – Opportunities in the Food and Agricultural Sectors,* Ripon, North Yorkshire.

immunogens feasible. Problems in immunisation remain, however, and it is important to determine the optimal method of inducing immunity. Novel methods of immunisation are being developed to exploit the ability of the immune system to produce antibodies against antibodies (anti-idiotype antibodies) which can then act as surrogate antigens where the natural antigen is difficult to isolate and purify, or is only weakly immunogenic.

c)   *Other therapeutic agents*

A third application of biotechnology is in the genetic modification of micro-organisms to design new antibiotics, for example to overcome bacterial resistance to existing products. There is hope that specific antibacterials and antiparasitic antibiotics may result from new research and that difficult fungal infections such as may occur in AIDS patients, mycobacterial diseases (tuberculosis and leprosy) and even certain tumours may be treatable by means of bioengineered antibiotics. The new antibiotic ivermectin is already proving very effective in treating external and internal parasites of companion and farm animals. Monoclonal antibody treatment is available against calf scours, and similar approaches are being developed for mastitis and coccidiosis in cattle. Synthetic interferons and various interleukins are being examined for their ability to combat the lowering of resistance to viruses and bacteria, as occurs in cattle shipping fever, and antibodies against interleukins in diseases which are due to the overproduction of such compounds.

d)   Animal and human diseases

Animal models are in great demand for elucidating the cause of human disease (e.g. Creutzfeldt Jakob disease, leprosy, carcinomas) and other species (foot-and-mouth in cattle). Genetically manipulated animals may prove invaluable for experimental investigations of the etiology of disease, as in the development of the oncomouse for the study of mammary tumours. Procedures for the introduction of disease resistance are also being investigated. The so-called major histocompatibility system is a part of the genome particularly concerned with resistance to disease, and work in farm animals should offer future prospects of breeding out disease susceptibilities. A good example is that of scrapie in sheep, where a simple blood test has been devised to detect whether or not the sheep possesses the gene that confers susceptibility to the disease. A similar test might be applicable to bovine spongioform encephalopathy (BSE or "mad cow disease") in cattle and to the allied chronic wasting disease in deer. Because of the resemblance to Creutzfeldt Jakob disease and Kuru in humans there have been fears that the agent responsible for BSE may be transmissible to man and that new prognostic procedures are urgently needed. Another example is the halothane gene in pigs which is associated with susceptibility to stress and low meat quality. A comparable condition of malignant hyperthermia has been identified recently in man so further study of the halothane gene linkage group will be of benefit for both species. Similarly, the discovery of a hereditary form of haemophilia A found in White Alpine sheep in Switzerland may prove to be an important model of human haemophilia as the clinical symptoms of extensive bleeding in various organs caused by a greatly diminished level of blood clotting factor VIII and the X-chromosome linkage of the disease, are features common to both species.

Diseases of farm livestock are the source of immense deprivation and economic hardship to the communities dependent upon them. They are also the cause of much animal suffering which all men of goodwill would wish to avoid. There is also an accompanying public health risk, where diseases are communicable to man or where animal products are contaminated with pathogenic organisms (see Chapter III). For all these reasons the opportunities provided by biotechnology to reduce or eliminate such hazards need earnest consideration. The new scourge of bovine spongioform encephalopathy is a good example of the type of unpredictable threat an advanced biotechnology capability should be able to cope with.

3.   Animal Production

a)   Animal growth and lactation

The World Bank estimates that food reserves represent only about 4 per cent of world needs, or about 15 days supply in the inconceivable event of global famine. An important component of the human diet is animal protein, which provides more than half

Table 13. **Impact of bovine somatotropin (bST) use on animal numbers, feed requirements, and waste production of dairy cows to achieve 1988 US milk production[1]**

Variable	Impact by bST[2]
Animals	
Cow numbers	decrease by 10.7%
Milk yield per cow	increase by 12.0%
Feed[3]	
Energy equivalent as corn grain	decrease by 2.5 million tons
Protein supplement equivalent as 44% soybean oil meal	decrease by 56 000 tons
Waste	
Manure[4]	decrease by 6 million tons
Urine[5]	decrease by 8 million m^3
Urinary nitrogen[5]	decrease by 80 000 tons
Methane[6]	decrease by 80 million m^3

1. US 1988 milk production values were 10.24 million cows, 6 460 kg milk per cow and 30 million tons total milk production.
2. Assumed 100% adoption and that use would increase average animal milk yield per cow by 12%. If commercially approved, expected impact would be less because technology rarely achieves 100% adoption.
3. Based on nutrient requirements for dairy cows averaging 650 kg body weight and producing milk of 3.5% fat content.
4. Based on an average diet composition of 1.62 Mcal net energy/kg and digestibility of 65%, and fecal dry matter of 16%.
5. Based on a daily urine production of 20 liters per cow with 1% nitrogen in urine.
6. Assumed that methane production represents 5% of gross energy intake.
Source: Bauman (1990).

the protein intake of adults. The ability of ruminant farm animals to utilise plant matter in their diet plays an important part not only in the exploitation of lowland and hill pastures but also in the economy of animal feed. Increased efficiency in food production from diminished acreages and enhancement of food quality have become important issues throughout the agricultural community. Animal feed is a costly and valuable resource and greater efficiency in protein accretion improves animal value and reduces inputs of expensive animal, plant and fish protein. Growth hormone (GH), which could be regarded as a ''nitrogen-fixation'' agent in mammals (Table 13), is known to improve milk production efficiency and carcass composition in farm animals. Efficiency is measured as the amount of milk produced for each kilogram of food eaten, whereas carcass composition refers to the amount of lean meat (protein) compared to the fat deposited. Since meat, milk and butter surpluses can be rapidly eroded, the need for increased efficiency will be even greater if the agricultural industry is to remain competitive and supply the needs of new and expanding markets in the face of reduced subsidies and decreased acreage under cultivation.

The potential of biotechnology is powerfully illustrated by the short history of recombinant bovine somatotropin (bST), which is the genetically engineered form of bovine growth hormone (bGH) naturally secreted by the pituitary gland. In the early 1980s the gene responsible for bGH production was isolated and transferred from animal to bacterial cells to produce large quantities of bST. Injections of about 30 mg/cow result in significant increases in milk production (about 10 to 20 per cent) which persist as long as treatment continues. Improvements in the efficiency of milk production mean that

Table 14. **Summary of progress of biotechnology-based somatotropins**

Material	Animal	Companies active	Status	Expected approval
Bovine somatotropin (bST)	Dairy cattle (lactation)	Monsanto/Genentech Cyanamid/Biotechnology Upjohn/Amgen IMC/Biogen Lilly	Large-scale field trials	After 1990
Porcine somatotropin (pST)	Pigs (growth)	IMC/Biogen Smith Kline/Amgen Cyanamid/Biotechnology General	Large-scale field trials	After 1990
Ovine somatotropin (oST)	Sheep (growth)	Amgen	R&D	After 1995
Avian somatotropin (aST)	Chickens (growth)	Amgen	R&D	After 1995

Source: Bloomfield (1988).

fewer animals will be required to produce the same volume of milk (Table 13). Controversy surrounds the adoption of bST in the dairy industry, not simply because of consumer concerns about the quality of dairy products from treated animals but also because of perceptions about the impact of the technology on small dairy farms and on the health and welfare of dairy animals (see Chapter VIII).

Related programmes with porcine somatotropin (pST) show that body fat can be reduced by up to 80 per cent and feed efficiency improved by 20 per cent. The use of GH in sheep and chickens is not so advanced, though two months of treatment increases growth rate in lambs by 36 per cent with a 30 per cent reduction in visceral fat. Clearly, the introduction of somatotropin would go some way to meet medical opinion which advocates a reduction in fat content of meat to reduce the risk of cardiac-related disease in man. Several major companies are active in the use of bST and have reached different stages in seeking approval for commercialisation. The present state of the bST biotechnology race is summarised in Table 14.

From extensive research, no evidence has emerged that normal doses of bST increase its concentration in milk, nor that bST alters the major constituents of milk such as fats and proteins, or increases metabolic disorders in dairy herds kept under a wide range of husbandry conditions. The parent compound is naturally produced in the body; it is a protein broken down by digestion in the gut or by the high temperatures of cooking; and it is inactive in humans since growth-retarded children do not respond to bST, but require the human homologue hST to produce a beneficial effect. Milk from cows treated with bST may have a higher concentration of a growth factor (insulin-like growth factor, IGF-1) normally present in cows' milk, though concentrations are similar to those in breast milk of women in early lactation and the growth factor is a normal constituent of human saliva. It is now believed that bST stimulates milk production in cows by raising the amount of the natural intermediate growth factor, IGF-1.

Other alternatives are being considered for the enhancement of animal performance which avoid regular injections. Vaccination against somatostatin, a hormone that inhibits release of GH, produces growth increments of 10-20% in cattle, sheep and pigs, though the response is less marked in high performance breeds. Certain monoclonal antibodies raised against GH enhance the biological activity of injected GH, though the mechanism is unknown. Antibodies against bST monoclonal antibodies result in bST-mimics which might improve the efficiency of growth or milk production and reduce the amount of fat present in the body. The bST-mimics last longer in circulation, possibly for months at a time, so that frequent treatment would be unnecessary. Immunisation against cell membranes from fat depots reduces the number of fat cells (adipocytes) in major fat depots in the body and provide yet another approach to the improvement of meat quality.

Detailed assessments have been carried out by major biotechnology companies to determine risks associated with the introduction of these various procedures and of bST in particular in different husbandry systems in various countries including Argentina, Kenya, Morocco and Venezuela (see also Chapter VIII). The value of bST in dairy cows farmed in the developing economies is also being investigated in Brazil and Zimbabwe. Results from some of these studies may help to establish whether a greater response occurs in genetically unselected animals. Lactating Murrah buffalo in India showed an average increase in milk yield of 13 to 24 per cent when treated with bST for 14 days. Public opinion in OECD countries, however, has remained sceptical partly because of the lack of informed comment about the rigorous evaluation of new products of biotechnology, and partly because some people think there has been too little academic research and development from sources independent of multinational enterprises.

b) Nutrition

Animal feed is a major expense of food production in OECD countries and one of the farmer's largest production costs. Biotechnology is predicted to have an impact on animal feeds through the production of feed additives including vitamins, enzymes and amino acids. Recombinant DNA technology may replace conventional microbial and chemical synthesis because with it microbial organisms can be manipulated and genes (e.g. coding for amino acids) transferred into more efficient systems. Investigations are in progress for the biotechnological production of amino acids such as lysine, which may improve the lean-to-fat ratio in meat production, and tryptophan, a limiting essential amino acid in pigs and poultry. Enzyme additives are in demand for biomass degradation and include glucose oxidase, beta-glucamase, cellulases and phytase. Pretreatment of forage crops with the enzymes cellulase and xylanase or with genetically-engineered organisms and the inoculation of silage are being evaluated for improvement of feed quality.

A 10 per cent improvement in forage digestibility could have a value to the UK dairy industry in excess of £10 million (sterling) per annum. Cellulase genes from rumen micro-organisms and thermophilic bacteria have been cloned and expressed in other bacteria. Inoculation of genetically engineered organisms into the rumen to facilitate cellulose digestion holds promise, and the introduction into rumen bacteria of synthetic genes coding for polypeptides of defined amino acid composition is being actively pursued, as are techniques for reducing methane production and protein degradation within the rumen but enhancing nitrogen fixation. Tissue-specific expression of cellulase genes in transgenic animals, e.g. targeted to the exocrine pancreas, suggests a novel approach to increasing the efficiency with which animals process food.

Probiotics offer an alternative approach to improving digestive function and are favoured by the desire for natural foods with no additives. Such diets consist of viable microbes added to the feed to increase the population of desirable organisms in the gastrointestinal tract (e.g. *lactobacilli*) at the expense of others causing disease. Studies in young pigs and sows have shown that probiotics reduce scours and improve weight gain and feed efficiency. Efficacy is also claimed in animals exposed to environmental stress or undesirable feed regime giving low resistance and proliferation of undesirable bacteria. However, the prospect of introducing probiotic microbes into the environment presents special problems that will require vigorous evaluation and strict quality control.

Many farm animals are fed antibiotics for the greater part of their lives. Pressure is growing to reduce antibiotic usage because of consumer hostility and the danger of resistant strains building up in the food chain. As mentioned previously, biotechnology provides an opportunity through gene cloning to develop new strains of antibiotics to compete with resistant ones.

Examples can be found where the application of biotechnology, though promising, has not been realised because of economic considerations. The potential of single cell protein (SCP) from yeast, bacteria or algae is one example since on economic grounds it is not viable as an animal feed supplement despite its high protein, essential amino acid, vitamin, mineral and carbohydrate content (see Chapter IV). Aerobic bacteria have been exploited using methane and ammonia as feedstocks to produce SCP but application of SCP in poultry, pig and fish feeds remains too expensive. At present, r-DNA techniques are being used to improve the production efficiency of bacteria, and time will tell whether this approach to the production of cheaper feed will ultimately prove acceptable.

4.   Animal Reproduction

Very few of the animal species with potentially useful attributes have actually been domesticated by man, and their genetic diversity has been further reduced by selective breeding for a small number of characters. Thus we now have breeds of farm animals that would never have arisen through natural selection. A universe of genetic characters eludes us that might help expand the utility of animals and their products in varied environmental conditions, including underdeveloped countries in tropical and subtropical regions of the globe. Biotechnology offers the means of speeding up the supply of desirable genomes which could be directed to particular needs, many perhaps dependent on unforeseen changes in climate, population dynamics, economics or social taste.

a)   Embryo multiplication

The established procedures of artificial insemination, cryogenic storage of semen, multiple ovulation and embryo transfer (MOET) in the bovine are being supplemented by twinning or multiplication of embryos. Twinning is achieved simply by bisecting the blastocyst through the inner cell mass, which is destined to become the embryo, and then transferring the reconstituted half embryos to recipient ewes or cows. A number of commercial embryo transfer companies offer this service, and some offer the prospect of implanting twin beef embryos into dairy cows to promote simultaneously both milk and beef production.

New approaches to embryo multiplication involve the culture of many identical embryos in the laboratory by one of two general methods. The first employs micromanipulation to separate the embryonic cells (blastomeres) from each other at, say, the four cell stage with onward culture and subdivision seriatim, until many embryos are obtained and stored in an embryo bank. Though conceptually simple, this approach has yet to be applied commercially. The more favoured approach is by nuclear transfer. All the genetic information necessary to construct a new individual resides in the cell nucleus, so provided a source of identical nuclei is available, ideally from animals of proven merit, their genetic blueprint can be imposed upon any ovum whose own nucleus has been removed and replaced by the donor nucleus. A virtually inexhaustible supply of immature ova can be extracted from slaughterhouse ovaries, and techniques have been designed to ripen these in culture in the laboratory. Donor nuclei can be obtained from cells of the inner cell mass in embryonic vesicles (blastocysts) or probably in future from embryonic stem cell lines maintained in culture. Animals of several species have been produced in small number by nuclear transfer but the technology is still at the laboratory stage of development, though it should have an impact on the livestock industry before the end of the decade.

It is important to realise that embryo multiplication adds nothing new to the genetic pool. The value lies in enabling rapid proliferation of desirable genomes, which might otherwise be confined to small numbers produced by artificial means, for example embryos formed by *in vitro* fertilization of cultured ova, and then sexed by fluorescent probes before being deposited in an embryo bank. Currently, DNA probes are available that selectively bind to the Y-chromosome specific DNA in male cells *in vitro*, and, in conjunction with powerful amplification techniques (polymerase chain reaction, PCR), allow an accurate diagnosis of the sex of an early embryo from one or two isolated cells *in vitro*.

At present the success rate of producing live offspring by transfer of embryos derived by nuclear transfer to female recipients is not high. Hence, embryo multiplication by blastomere separation as described above would be essential to maximise the beneficial consequences of a new animal variety produced by transgenic technology, which might include more efficient feed utilization, disease resistance, and improved quality of meat, milk, fleece, leather or other products. The development of techniques that would give multiple copies of a particularly valuable adult animal are still a long way off. It would require the transfer of somatic cell nuclei from that animal to enucleated ova to produce genetically identical, cloned offspring. Scientists do not yet know how to reprogramme the nucleus of a somatic cell so that it becomes like that of the fertilised germ cell with the capacity to form a new individual. This remains one of the fundamental problems of biology.

The new options that such combined biotechnologies could offer would be the large-scale production of animal products of predetermined specification to supply established markets, or to meet national or acquired needs, or to bring fresh breeding lines of known provenance to upgrade flocks and herds in conventional breeding programmes. An additional option that may assume importance in the near future would be to use embryo multiplication technology and embryo banks to offset fears of shrinking gene pools arising from increasing breed specialisation in agriculture. This could be achieved by expanding the existing numbers of rare breeds and feral stocks of cattle, sheep, pigs and poultry and preserving their genomes in cryogenic embryo banks as a resource of unexploited genes for an unpredictable future. Qualities such as coat colour and texture, body size and carcass fat distribution, meat flavour, fertility and ease of parturition, work capacity, hardiness and phenotypic variation are commonly more distinctive in rare breeds or feral stock than in conventional farm livestock. Any or all of these characteristics could become prized, for example to promote extensive livestock production or to diversify environments and livestock variety.

b)   Gene transfer

Transgenic technology, whereby gene sequences manufactured by recombinant DNA techniques are introduced into animal embryos by pronuclear microinjection, retroviral vectors or possibly by transfer by spermatozoa, will eventually provide the capability to generate new characters or products (Table 15). The best publicised examples are the insertion of growth hormone genes into pig and sheep embryos to increase leanness and growth rates after birth, the breeding of transgenic chickens resistant to Marek's disease and of transgenic sheep secreting human Factor IX (for treatment of haemophiliacs) in their milk. In studies of growth, giant mice have been produced by injecting a human GH gene into fertilised mouse eggs. Less spectacular results have been achieved so far in domestic farm animals using a similar gene construct or in poultry using a retrovirus to transport modified DNA into the genome of the host. In pigs only about 1 per cent of embryos injected with human GH resulted in transgenic offspring. Growth rate was rarely enhanced though carcass fat was sometimes reduced, but the increased incidence of arthritic lesions, lameness and infertility has raised serious doubts about this approach. Recent work, however, has shown that transgenic pigs with a bovine GH gene regulated by a different promoter sequence (for prolactin) are healthy, have a normal pattern of reproduction and show episodic release of bGH in response to suitable extrinsic stimuli. Moreover, transmission of the foreign gene from parent to offspring occurs at the expected frequency.

Transgenic animals have also been prepared with gene constructs that code for the production of human alpha-1 trypsin (useful in certain respiratory disorders and acute shock in man) and human interleukin-2, a regulator of immune cell production. By linking the gene to sequences from another that is normally expressed in the mammary gland, selective expression has been achieved in the gland and in no other tissue. This is a form of "targeted expression" with the gene product being formed in a new site from which it is readily recoverable without producing any abnormalities. For those alarmed by the idea of transferring "human" genes to other animals, it is important to note that many genes can now be manufactured in a gene synthesizer (oligonucleotide synthesis), so that accurate copies can be produced. When introduced into the early embryo the copy of the foreign gene becomes but one of more than 50 000 genes contained in every cell in the body.

Other potential applications include the augmentation of cysteine content to enhance wool quality in sheep, genetically manipulated casein genes to enhance the cheesemaking characteristics of milk, or alpha-lactalbumin genes to reduce the lactose content of cow's milk and render it more assimilable by humans with lactose intolerance. Various genetic switches, or promoter sequences, are being studied that improve targeting gene expression to muscle (e.g. actine promoter linked with insulin-like growth factor-1) or facilitate responses to changes in environmental temperatures (e.g. heat shock protein promoter

Table 15. **Development of transgenic animals**

Feature	Description	Comment
Novel Step	Transgenic animals	This is a general technique for animal transformation by inserting useful genes
Commercial Consequence	Varies with application	Prospects are: a: improved performance (growth, food conversion, disease resistance); b: improved animal composition (lean meat, wool quality, milk); c: new products (high-value proteins secreted in milk/ blood); and d: models for medical research
Current Market and Technology	Traditional animal breeding. Pharmaceutical products	First transgenic product is feasible: a blood product for treating haemophiliacs
Major Benefits	This technology will revolutionise livestock industry	Careful choice of developments is needed. High testing costs
Commercial Infrastructure	Highly organised in developed world, especially poultry, pig, animal breeding and pharmaceutical industry	Developing countries more diverse, and disease resistance a high priority
Sectors Affected	Animal breeding and production. Feedstuffs. Pharmaceutical. Medical. Animal health. Dairy. Meat	Among many others
Next Technical Steps	Identify useful genes. Isolate and clone. Obtain controlled expression. Test	Gene insertion is available. Depending on market, safety, efficacy and side-effect testing, may be very expensive
R&D Base	Very significant world-wide	But many separate opportunities
Time and Cost to Market	Varies with application	Unlikely to be less than 3 years and £1.0-2.0 million
Competing Technology	Conventional breeding. Mammalian cell culture	None for some applications

Source: Prospect Management Services (1988), *Biotechnology for Businessmen – Opportunities in the Food and Agricultural Sectors,* Ripon, North Yorkshire.

linked with growth hormone). The prospect of recovering substantial quantities of pharmaceutically valuable proteins in active form from milk or saliva of sheep, goats or cows uncontaminated with human viruses (such as HIV), and in amounts greater than those obtained in culture, has been recognised as having important medical implications.

Current methods of producing transgenic animals are difficult to execute and uncertain in effect, so much more work is required both to identify the useful genes and regulatory DNA sequences and to determine the optimal conditions for their introduction and expression. In respect of avian biotechnology the potential for improving the performance of the domestic fowl by genetic modification has been recognised for almost a decade, but it has proved impracticable to introduce manipulated DNA into the oocyte nucleus. Alternative approaches are being investigated, including injection of DNA into the primitive disc of embryonic cells, or forced incorporation of DNA into embryonic cells grown in tissue culture with subsequent inoculation of transfected cells ("infected" by genes) into recipient embryos. Model experiments have been carried out with non-replicating retrovirus vectors to demonstrate that foreign DNA can be introduced into the chicken embryo by this means, as in mammals. If some of the transfected cells come to reside in the gonad then transgenic progeny may subsequently be produced. This is known as germline transmission. Further advances in these techniques have shown that it is possible to use a replication-defective vector to transfer heritable non-viral genetic information into the chicken germline. However, knowledge that transgenic birds have been produced in this way might be expected to provoke a negative response, despite the fact that people have been eating poultry meat that frequently contains retroviral sequences. Extensive evaluation of these new procedures would therefore prove obligatory.

Production characteristics seem to be controlled by interactions between many genes, but the aim of introducing disease resistance is to identify genes that have a major influence. In chickens, the immunoglobulin heavy chain and the light chain genes could have relevance to disease resistance, but it is considered more appropriate to search for genes which give resistance to specific pathogens, since the correlation between immunological responsiveness and disease resistance is weak. Transgenic chickens resistant to avian leukosis virus have been produced by the stable introduction of a DNA copy of the virus genome.

Experimental studies in transgenic biology provide insights into the regulation of gene expression for the improvement not only animal efficiency but also animal health and welfare. Special attention is being given to understanding how a gene is switched on or off at various times in development and in particular tissues of the body. Such is the potential of molecular biology that some of these questions are now being answered. Few would raise doubts about techniques which seek to improve the quality of life of animals kept under different forms of animal husbandry. However, strict legislation has been introduced in various OECD countries to ensure that reproductive technologies for genetic modification, cloning and hybrid production should be confined to animal populations and that application to human embryos should be ruled unethical and a criminal offence. The number of occasions when gene transfer could be legitimately contemplated for embryos of agriculturally important animals is so far very limited. In many instances, the identity of important genes that regulate animal performance and disease resistance is not known.

The pace of developmental biology is so rapid that discoveries in chimaerism, parthenogenesis, and stem cell biology may provide still more opportunities for regulating the genetic and physical form of domesticated farm animals. For instance, embryonic stem cells can be cultured in such a way that they retain their capacity for self-renewal but are restrained from differentiating into defined cell types. The undifferentiated cells can be cultured over the long term, providing the opportunity to introduce new genes that enhance performance or to repair faulty ones that impair function. This new advance means that genes can be introduced into these cells *in vitro* so that precise genetic modifications can be established and confirmed before the stem cells are reintroduced into the early embryo. Therefore, embryonic stem cells will become the method of choice in the future for improving the genetic quality of farm livestock because they will be more efficient than oocyte microinjection for the introduction of genes into the germline.

c) Genome analysis

In view of the importance of farm animals to mankind's needs in respect of nutrition, clothing, transport, traction and leisure activities we must continue to work for improvements in these attributes for the benefit of future generations. Scientific work on farm animal genetics aims to understand the real nature of the genome and the principles of inheritance. In future, gene mapping is likely to become increasingly important not only in the elucidation of theoretical principles that apply between breeds and species but also in practical ways for the improvement of animal breeding programmes.

Within the genome, the linear arranged information on DNA can be influenced or changed by events occurring during meiosis, such as crossing over between chromosomes and segregation of genes. Mutational events are rare but important, and it is becoming increasingly necessary to understand how genes are regulated under different environmental conditions within various tissues or organs and in whole organisms or populations. Molecular and cell biology now provides new tools to attack these problems, including cell hybridisation, specific identification of gene products and how they vary in nature, detailed analysis of overlapping segments of DNA, investigation of recombination frequencies in family studies and the use of so-called minisatellite and microsatellite techniques to detect polymorphic segments by which informed selections can be made.

Present information on gene assignments in farm animal species is rudimentary, and an expansion of investigations into where specific genes are located on bovine, porcine, ovine and avian chromosomes is urgently needed. These studies are important not only in terms of comparative chromosome organisation, but primarily as a means of mapping the loci that affect production characters, disease susceptibility or disease resistance. Progress here would also improve the prospect of manipulating these loci in breeding programmes at a genotypic rather than a phenotypic level. For cattle and pigs, methods outlined above and developed to map the human genome are being applied and a European Community BRIDGE programme has been initiated called PigMap. These methods will provide somatic cell hybrids containing specific chromosomes and enable the localisation of genes on identified chromosomes and linkage analysis to determine whether individual loci are on the same chromosome. It is believed that a bovine and porcine gene map consisting of a few highly informative marker loci per chromosome is an achievable goal. Although present knowledge in these two species is ten years behind that of the human genome, today's powerful methods are rapidly narrowing the gap. It is becoming urgent to devise appropriate procedures to allow access to developing data bases arising from the international human genome projects.

5. Fish

a) Aquaculture

In contrast to conventional livestock production, fish farming is a new area of agriculture with the fish under cultivation still relatively wild animals. Although technology over the past decade has provided some improvements in productivity, much of the world's aquacultural production of some 15 million tonnes, which is approximately one-fifth of the total fish harvest, is produced with little assistance from the scientist. Since 85 per cent of the global surface is water, the scope for technological improvements on current production is very great. Given the ecological constraints, the increasing difficulties of catching or trawling wild fish stocks and the increasing recognition of the nutritional and medical advantages of fish in the diet, one must conclude that aquacultural production is certain to grow significantly.

Fish have many advantages over land animals. Being cold-blooded they do not have to expend energy in maintaining high body temperatures; their aquatic milieu removes the need for energy expenditure against gravity. Consequently, fish are able to provide better food conversions than most other farm animals. They also live in a three-dimensional "field" and thus more effectively use available space.

The range of fish species available for modification is also very large, larger in fact than all other vertebrates combined. This gives considerable opportunities for exploitation of different ecological niches and cultivation conditions. Their genotypes are relatively simple, their fecundity or egg-producing capacity is very large and their sex or gender control malleable, all factors which make them eminently suitable for technological modification. Their relatively short history in culture and the lack of genetic selection compared to higher domesticated animals means that scope for stock improvement is much greater.

Successful agricultural production, of which aquaculture is clearly an important and fast developing component, is dependent on our understanding and control of disease, nutrition, genetics and reproduction. Already biotechnology has assisted production in these areas. Sex reversal and the environmental and hormonal induction of spawning, growth performance enhancement, antibiotic and vaccine development, parthenogenesis and ploidy manipulation are all techniques that are being used to raise productivity. Many of these techniques and the processes that are being modified are unique to fish.

Further quantum advances are to be expected with the application to aquatic organisms of the recombinant DNA technology currently available for higher organisms. This section provides a brief survey of the way that biotechnology is currently influencing or is about to influence fish production. Recombinant DNA technology is beginning to make an impact on aquaculture, at least at the research level, as witnessed by the increasing proportion of papers devoted to such topics in aquaculture genetics conferences since 1987/88. The main areas of interest are transgenic fish, basic molecular genetics (gene expression and cloning), population genetics and pedigree analysis. However, it is only the latter two which have so far had any influence on husbandry or production, and even then only in terms of the high-value cold water species.

b) Reproduction

In aquaculture, the ability to control the reproductive cycle of the species under cultivation is of paramount importance. Induced spawning enables supplies of eggs and fry to be made available from fish which would not spawn naturally in captivity. For species which are able to mature and spawn under farm conditions, controlled reproduction can provide seed supplies at precisely those times required by on-growing farms and not just during the few weeks of the year when natural spawning occurs. The injection of gravid fish with crude extracts of pituitary gland (called hypophysation) has been used to induce spawning in carps and other species for many years. New developments with synthetic hypothalamic peptides have improved spawning induction. However, further understanding of the importance of environmental conditions in causing maturation has led to the development of spawning controls in salmonid fish without resorting to a treatment with exogenous hormones. Photoperiod manipulation of broodstock to ensure a year-round supply of eggs is widely used in the rainbow trout industry in North America and Europe.

In addition to being able to produce good quality seed on demand, it is also essential that aquaculturists be able to prevent or inhibit sexual maturation. During maturation many fish undergo profound changes in growth and flesh composition which at worst render the flesh unacceptable for market sale and at best offer poorer performance from the stock and wide variation in flesh quality. Methods of sex control involving either sex reversal or sterilization are able to ensure the channelling of the feed eaten into growth rather than the demands of maturation. Gender is extremely plastic in fish and direct treatment of fry by short-term feeding or immersion can induce sex reversal so as to produce either male or female individuals as desired. Because of concerns about consumption of hormone-treated fish, the technique is used to provide sex-reversed individuals of the homozygous sex as broodstock. By crossing to normal homozygous individuals, monosex fry are produced in the next generation.

In aquatic organisms the scientist has the ability to modify the number of chromosome sets inherited and their parental origin by relatively straightforward physical (pressure, heat, cold) and chemical (cytochalasin B) treatments.

Gynogenesis and androgenesis are genetic modifications whereby individual organisms inherit genes only from the female or male parent respectively. Generally they are achieved by destroying the chromosomal material of one gamete and then preventing either the second meiotic division or first cell division to restore the chromosome number (diploidy). Gynogenesis has already been used for creating inbred lines in a relatively short period, and in studying sex determination. It is currently impossible to cryopreserve (store successfully at very low temperatures) the eggs or embryos of finfish species, which would be useful for "gene bank" purposes. However, cryopreserved sperm can be used to produce diploid individuals by androgenesis (irradiation of the unfertilized donor egg followed by suppression of the first cell division to diploidize the haploid sperm genome). Androgenetics and gynogenetics produced by suppression of the first cell division are completely homozygous (i.e. completely inbred) and offer ways of studying clonal lines and levels of inbreeding which are generally unobtainable in other agricultural animals.

Similar ploidy manipulation techniques are already used to produce polyploids in many aquatic organisms (trout, carp, salmon). The use of all-female triploid fish (having three sets of chromosomes) which are effectively sterile and are free of maturation problems (which cause considerable financial losses to the farmer) is now widespread in the salmon and trout industry. Their sterility also prevents escapees from genetically polluting geographically isolated stocks in nearby wild fisheries. The survival of certain hybrid crosses is greatly improved if they are made triploid, and this also greatly extends the combinations of species which may be crossed and thus the opportunities to make use of the genetic diversity found in aquatic organisms.

Early work on the production of tetraploid strains is promising, and some have proved to be fertile. Crosses between tetraploids (which have four sets of chromosomes) and normal female strains of rainbow trout have also produced triploid offspring. This removes the requirement for treating every individual egg, as is necessary in the triploidy production techniques now available commercially. It is important to carefully monitor these methods, and molecular genetics now offers a much broader range of inheritance markers than the few available in most species using traditional genetics.

Similar techniques are also being applied to organisms other than fish which are used in aquaculture. Triploidy induction seems to be widely applicable in molluscan species, using similar techniques to interfere with meiosis in fertilized eggs. Oysters are being used as a model species to develop cryopreservation of eggs and embryos in aquaculture. The large number of eggs produced by many species used in aquaculture makes the cryopreservation of these life stages an attractive prospect, but while sperm cryopreservation is fairly routine, there are major problems for most species due to the size of the eggs. Oyster eggs are small enough to allow successful development of cryopreservation techniques which may help discover methods for larger eggs and embryos.

The various techniques described above have already had a significant impact on aquaculture, as in the regulation of the sex determination system of rainbow trout to produce all-female stocks, which is widespread and has apparently met with no consumer resistance. Likewise triploid rainbow trout (which are sterile and thus marketable during the spawning season) are also sold. The use of such techniques in the salmon industry may meet with greater resistance because of the marketing image of farmed salmon which are presented as being virtually the same as wild fish.

c) *Transgenesis*

Some techniques for introduction of cloned DNA into fertilized fish eggs have been developed for a number of fish species. In most cases they have involved microinjection of relatively large amounts of DNA (106 to 108 molecules) into the cytoplasm of fertilized eggs before the first cell division. In general, it has not proved possible to inject DNA into nuclei due to an inability to visualise them in fish eggs. Despite these difficulties of injection, there is good evidence for both the integration (i.e. incorporation into the chromosomal genetic material) and inheritance of genes introduced into fish ova. Expression of introduced genes has also been observed, in some cases without integration of the injected DNA.

Like many such experiments on agricultural mammals, interest in transgenic fish for aquaculture was largely stimulated by the increased growth rate of transgenic mice carrying metallothionein - growth hormone (MT-GH) fusion genes. The lack of suitable cloned piscine genes was an obvious drawback for such research in fish. There has now, however, been a considerable increase in the number of laboratories both cloning fish genes and studying basic aspects of fish molecular genetics. As might be expected, increasing growth rate has been one of the prime objectives. Growth hormones from a variety of salmonid species as well as the rainbow trout metallothionein gene have now been cloned.

A variety of purified growth hormones will accelerate growth when injected into fish. Although the mammalian metallothionein promotor functions at a much lower level in fish cells than the endogenous piscine gene, expression of growth hormone from a mammalian MT-GH fusion gene has been detected in fish eggs. An increase in growth in transgenics carrying such constructs has also been reported, though without direct evidence of growth hormone production. However, other experiments have failed to detect expression from such constructs. Carp which are transgenic for a rainbow trout growth hormone sequence have been reported to grow faster than non-transgenic siblings and have been the subject of the first outdoor trial of transgenic fish in the United States. These fish were injected with a trout growth hormone gene linked to a constitutive promoter (a regulatory DNA sequence which causes continual expression of the gene it controls). This was the first example of integration, expression and inheritance of introduced DNA in experiments on transgenic fish.

Following this initial phase, it is to be hoped that the broader establishment of research on applied molecular genetics of fish will lead to success not only in areas such as increasing growth rate but in other aspects related specifically to organisms used in aquaculture. One potential example is the attempt to increase cold tolerance in Atlantic salmon by transferring winter flounder antifreeze protein (AFP) genes into the genome of the salmon. It is hoped that expression of this protein in the transgenic fish will allow their survival in the north of Canada, where winter sea temperatures preclude salmon culture at present.

Due to the short history of such techniques in aquaculture research, it is difficult to predict which directions will be taken in the future. The use of antisense genes may have potential in the control of certain viral diseases of fish, although such therapeutic applications of these methods are a long way off. The development of mass transfer techniques for introducing DNA into fish eggs or sperm, using methods such as electroporation rather than relying on individual microinjection of each egg, would facilitate the application of such a treatment on a large scale. Mass transfer would be of great use not only for fish gametes but for any organism producing large numbers of eggs and having external fertilisation. This includes the vast majority of vertebrate and invertebrate species used in aquaculture. Such organisms could also prove to be valuable models for the development of methods applicable to other animals where there are problems such as low numbers of eggs or difficulty of access to newly fertilized eggs (e.g. birds).

Unlike most other agricultural crops, the products of aquaculture face direct competition from wild-caught sources, often of the same species. This introduces an additional factor in determining the economic feasibility of using relatively expensive techniques such as gene transfer in aquaculture. However, it does appear that transgenic fish (most probably sterile and possessing increased growth rate potential) will be ready for full-scale commercial testing in the near future. Such possibilities have raised fears about future monopoly in aquaculture by large companies, particularly in developing countries, and at present several OECD countries have legislation in place banning release of transgenic animals.

As many stocks for aquaculture are still being drawn directly from wild populations, knowledge of population structure is of great importance in decisions about foundation stocks for new husbandry systems. The analysis of mitochondrial DNA polymorphisms for restriction enzyme sites has given a powerful new technique to population geneticists to add to the more traditional isoenzyme variation.

DNA fingerprints and restriction fragment length polymorphism (RFLP) analysis both have potential in pedigree analysis. They also have value as aids in verifying specialised genetic modifications such as gynogenesis (see above) which are used in aquaculture research. In an aquatic environment, precise monitoring of matings is not always possible, particularly where the biology of the organisms does not make *in vitro* fertilisation or controlled pair matings economically feasible for large scale breeding programmes. Monitoring of gene flow via both traditional and molecular techniques is of value in designing appropriate breeding schemes and preventing unwanted inbreeding.

d) *Disease*

Fish diseases are just as varied and as prevalent as those occurring in any other intensively cultured species, but the small size and low value of individual fish and the nature of their environment generally make clinical medicine and individual animal treatment impossible. The situation is further complicated by the nature of the aquatic environment, which makes external therapy very difficult and oral treatment only possible with fish that are actively eating. Environmental considerations, perceived or genuine, add further constraints to the approaches to fish and shellfish therapy.

Thus, there is now great emphasis on biological or immunological methods of control of fish diseases. Already there are several vaccines available against bacterial diseases, but these are generally relatively straightforward bacterins. Certain diseases, however, and most notably furunculosis, where the causative agent produces an immunosuppressive exotoxin, have not responded to such straightforward vaccinal approaches. Here, for example, efforts are being made to separate the lethal toxin that kills the fish from the suppressor that prevents an immune response, and then to develop large-scale production of the former by incorporation of its genome in yeast, for use as an antigen for vaccine production. Similarly, attempts are being made to produce by biotechnological methods large volumes of oral attachment or digestive enzyme antigens of external parasite mouth parts with which to induce mucosal antibody protection.

Monoclonal antibody production has been developed against a number of fish pathogens and used widely for research purposes. Diagnostic kits incorporating them are available, but they are not generally of great value. This is because the agents of most fish diseases are carried widely by non-clinically affected populations; although the sensitivity of such kits can be adjusted, stress is the major precipitating factor, and disease can often be induced by very low levels of infection.

Alternative possibilities with monoclonal antibodies lie in the use of blocking serum therapy in high-value broodstock at spawning time to prevent the incorporation of pathogens transmitted via the ovaries into the egg at ovulation and treatment with antibodies to inhibitory hormones for the induction of ovulation in difficult species.

Detection of the DNA or RNA of pathogens or their products in minute quantities within the tissues of carrier or affected fish by using DNA or RNA probes is an even more sensitive technique than the use of monoclonal antibodies. This technology has already been used on occasion for research purposes, and it can be expected to have wider use in disease research in the future. However, except for certification of disease-free status or for quarantine purposes, its diagnostic value, like that of monoclonal technology, seems likely to be limited by the ubiquitous nature of most fish pathogens.

6. The Future and Public Attitudes

It is a striking fact that few of the developments described in this chapter were envisaged ten years ago, despite the fundamental advances in molecular biology which had made them possible. New knowledge is a necessary but not a sufficient cause of technological progress. Nevertheless, the relevance of biotechnology in animal husbandry is now obvious from the number of commercial products which have already reached the market. Many more are due to become available soon, and it is predicted that biotechnology-derived products will comprise at least 30 per cent of the animal health market and services by 1995. A survey by the US Food and Drug Administration (FDA) in 1985 and 1986 showed that of 171 separate veterinary biotechnology projects, one-half were concerned with disease diagnosis, prevention and treatment, over one-third with nutrition, growth and reproduction and about one-tenth with feed additives. Growth hormones and vaccines are viewed as the most likely commercial prospects during the next five years.

During the 1980s there came about a remarkable degree of consensus on agricultural priorities in the developed nations, and this consensus bore little resemblance to that of the 1970s where higher outputs at almost any cost were the top priority. Now the call is for environmentally friendly production, which may be interpreted as less intensive animal systems, improved health and welfare of livestock, and lowered input costs of chemical fertilizers and herbicides. Public concern about release of genetically engineered organisms suggests that monoclonal antibodies for diagnostic kits and veterinary therapeutics will continue to reach the market first. Many small biotechnology companies have formed strategic alliances with universities or research institutes to provide access to research and new developments.

Without doubt, countries of eastern Europe will now wish to develop their agricultural systems to enhance home food production. With educated people and newly released energies, they will seek access to the biotechnological options already being created in the West for enhanced efficiency of meat, milk and egg production. There may also be special cultural needs in some countries not yet addressed by existing western technology, and new bioengineering procedures may be called for (e.g. to permit traditional cheesemaking, or natural fibre production). Subject to the necessary financial support, it is fairly certain that most of the options for more efficient animal health and production already developed in the West will be in demand to boost food production.

In developing countries, however, the problems are of a different kind. Pastoral or fishing communities devoid of industrial infrastructure cannot be expected to create their own biotechnology. Nor would they wish continually to be passive recipients of end-products in the form of famine relief shipments. Perhaps an effective solution to long-term agricultural improvement would include a combination of training of farmers in the pastoral villages and assistance with the genetic improvement and veterinary care of native livestock, exploiting the embryo production and transfer procedures and vaccines developed in Europe and America. Any such programme would require co-ordinated help with fodder crops and a considerable financial investment. To be sustainable, the whole-hearted commitment of the local population would be essential, together with minimum reliance on *in situ* technology.

Given the widely differing needs of advanced industrial nations, emerging eastern European democracies and the disparate countries of South America, Africa and Asia, the challenges to agricultural biotechnology are immense. Though the application of newly discovered biological principles takes time to translate into efficient delivery of vital commodities, it seems that biotechnology has the potential to develop fast enough to meet the needs of expanding world populations.

Issues remain, however, about consumer perception and consumer acceptability of biotechnology. In the field of human health, new products of biotechnology have already reached the market and offer medical healthcare in fields previously plagued by problems

Table 16. **Summary of major research opportunities and potential exploitation**

	Research Area	Potential	Exploitation	Approx. time scale (years)
1.	Growth and lactation	Transgenesis (gene expression and regulation)	New breeds with rapid growth potential and improved carcass quality (sheep, cattle, poultry). Increased efficiency of milk production (cattle).	> 5
		Receptor structure	Pharmaceutically valuable proteins.	< 5
			New potent ligands for activation of specific receptors and signal transduction.	< 5
		Immunoregulation (idiotypic and anti-idiotypic immunisation)	Stimulation of exponential growth phase. Regulation of fat cells.	> 5
2.	Reproduction	Stem cell biology and transgenesis	Gene transfer of commercially valuable genes.	5-10
			Multiplication of genetically superior animals.	5
			Disease resistance.	5-10
		Chromosomal analysis (gene mapping)	Sex selection. Selection of valuable traits.	< 5
		Growth factors and lymphokines	Reduce embryonic loss (10-30% in large animals). Use of interferons to overcome pregnancy loss.	5
3.	Diagnostics	Biosensors	New kits for detection of oestrus, fertility tests and parturition predictors (on-farm). On-line monitoring in dairy (fertility, mastitis).	< 5
			Sperm fertilizing ability.	< 5
		Disease	Wide range of applications for diagnosis of disease state or large-scale screening.	< 5
4.	Therapeutic drugs and vaccines	Novel antigen sources (recombinant antigens)	Parasitic disease.	< 5
		Antigen presentation (by immune cells)	Universal importance to development of new therapeutic procedures.	< 5
		Immunogenic vectors	Wide range of disease states.	< 5
		Immunisation	Anti-idiotypic (surrogate antigen) immunisation against pathogens.	> 5
		Disease resistance using vaccines	Important in developed and less developed countries	> 5

Amino acids	Feed additive to improve dietary balance.	< 5
Antibiotics	Novel antibiotics that overcome resistance.	< 5
Rumen micro-organisms	Cellulase enzymes for improved feed usage, silage fermentation and cellulose breakdown.	< 5
Intestinal micro-organism interactions	Probiotics.	5
Transgenic micro-organisms	Enhanced ruminal digestion.	> 5

of quality control (e.g. somatotropin) or a lack of suitable therapy (e.g. hepatitis B vaccine). In these respects, the consumer already has the benefit of an expanded pharmacopoeia which provides for greater healthcare as a direct result of biotechnology. Examples of future opportunities of animal biotechnology are given in Table 16.

In the present context, it is noteworthy that a recent report from the National Consumer Council of the UK raised three questions about biotechnology, is it safe?, is it fair?, and is it natural? The question of safety has been alluded to already. In respect of the fairness of the procedure, problems related to animal welfare exist for a compound that has to be injected daily into animals, as with insulin for diabetes in man. Oral administration is difficult because of gastrointestinal degradation; transdermal delivery may not be appropriate for larger molecules, and short half-lives in the blood circulation often enhance rather than resolve the problem. Such problems may dominate research and development in biotechnology companies, though major advance is just as likely from basic research laboratories in membrane transport, polymer science and enzyme kinetics. As to the question of whether techniques to change hormone concentrations in animals are natural, it is interesting to note that king poodles contain much higher concentrations of IGF-1 than miniature poodles, which have higher levels than toy poodles. Dog breeders appear to have selected naturally and unknowingly for the gene coding for IGF-1 over many generations! Perhaps what is considered natural is in reality merely what has become familiar.

7. Further Reading

BAUMAN, D.E. (1990), "Bovine somatotropin: review of an emerging animal technology", Office of Technology Assessment, Congress of the United States, Washington DC.

BLOOMFIELD, G. (1988), *Trends in Veterinary Research and Development,* Vol. 1, "An Overview", PJB Publications Ltd, 18/20 Hill Rise, Richmond, Surrey, UK.

BLOOMFIELD, G. (1990a), *Trends in Veterinary Research and Development,* Vol. 2, "Transgenic Animals", PJB Publications Ltd., 18/20 Hill Rise, Richmond, Surrey, UK.

BLOOMFIELD, G. (1990b), *Trends in Veterinary Research and Development,* Vol. 3, "Diagnostics", PJB Publications Ltd., 18/20 Hill Rise, Richmond, Surrey, UK.

BYE, V.J. and R.F. LINCOLN (1986), "Commercial methods for the control of sexual maturation in rainbow trout *(salmo gairdneri R.)*", *Aquaculture,* Vol. 57, pp. 299-309.

CRITTENDEN, L.B. and D.W. SALTER (1990), "Transgenic chickens resistant to avian leukosis virus", *Proceedings 4th World Congress on Genetics Applied to Livestock Production,* Vol. 16, pp 453-456.

CROSS, B.A. (1989), "Animal biotechnology", *Philosophical Transactions of the Royal Society B,* Vol. 324, pp. 563-575.

FERRIS, S.D. and W.J. BERG (1987), "The utility of mitochondrial DNA in fish genetics and fishery management", in *Population Genetics and Fishery Management,* eds. Ryman and Utter, Washington Sea Grant Program, Washington Press, pp. 277-300.

FRIES, R., J.S. BECKMAN, M. GEORGES, M. SOLLER and J. WOMACK (1989), "The bovine gene map", *Animal Genetics,* Vol. 20, pp. 3-29.

FRIES, R., P. VOGELI and G. STRANZINGER (1990), "Gene mapping in the pig", in *Domestic Animal Cytogenetics,* ed. R.A. McFealy, Academic Press, New York.

GELDERMANN, H. and F. ELLENDORFF, eds. (1990), *Genome Analysis in Domestic Animals,* VCH (UK) Ltd, Cambridge.

HALVER, J. (1989), *Fish Nutrition,* Academic Press, London.

KALTER, R.J., R. MILLIGAN, W. LESSER, W. McGRATH, L. TAUER and D. BAUMAN (1985) *Biotechnology and the Dairy Industry,* Cornell University Press, Ithaca, New York.

OZATO, K., K. INOUE and Y. WAKAMATOU (1989), "Transgenic fish: biological and technical problems", *Zoological Science (Japan),* Vol. 6, pp. 445-457.

PALMITER, R.D., R.L. BRINSTER, R.E. HAMMER, M.E. TRUMBAUER, M.G. ROSENFELD, N.C. BIRNBERG and R.M. EVANS (1982), "Dramatic growth of mice that develop from eggs microinjected with metallothionein-growth hormone fusion genes", *Nature* (London), Vol. 300, pp. 611-615.

PHIPPS, R.H., C. MADAKADZE, T. MUTSVANGWA, D.L. HARD and G. de KERCHOVE (1991), "Use of bovine somatotropin in the tropics: the effect of sometribove (USAN) on milk production of *Bos indicus,* dairy crossbred and *Bos taurus* cows in Zimbabwe", *Journal of Agricultural Science,* Vol. 117, Part 2, pp. 257-264.

POLGE, E.J.C., S.C. BARTON, M.A.H. SURANI, J.R. MILLER, T. WAGNER, F. ROTTMAN, S.A. CAMPER, K. ELSOME, A.J. DAVIS, J.A. GOODE, G.R. FOXCROFT and R.B. HEA (1989), "Induced Expression of a Bovine Growth Hormone Construct in Transgenic Pigs", in *Biotechnology in Growth Regulation,* eds. R.B. Heap, C.G. Prosser and G.E. Lamming, Butterworths, London, pp. 189-199.

Prospect Management Services (1988), *Biotechnology for Businessmen – Opportunities in the Food and Agriculture Sectors,* Ripon, North Yorkshire.

PURSEL, V.G., E.E. REXROAD, I.D.J. BOLT, K.F. MILLER, R.J. WALL, R.E. HAMMER, C.A. PINKERT, R.D. PALMITER and R.L. BRINSTER (1987), "Progress on Gene Transfer in Farm Animals", *Veterinary Immunology and Immunopathology,* Vol. 17, pp. 303-312.

SIMONS, J.P., I. WILMUT, A.J. CLARK, A.L. ARCHIBALD, J.O. BISHOP and R. LATHE (1988), "Gene transfer into sheep", *Biotechnology,* Vol. 6, pp. 179-183.

STRAUGHAN, R. (1989), *The Social and Ethical Issues for Consumers: A Discussion,* National Consumer Council, 20 Grosvenor Gardens, London.

WILLHAM, R.L. (1990), "Humanity and Livestock", in *Proceedings of the 4th World Congress on Genetics Applied to Livestock Production,* Vol. XIII, pp. 15-25.

Chapter III

FOOD

1. Introduction

This chapter considers the likely spread of biotechnology in the food and drink industries and in associated suppliers of ingredients and specialised services to the industry.

Many biotechnology developments are already available for application in foods. We begin, therefore, with a consideration of the factors which may encourage or restrain the commercial introduction of biotechnology processes, products or services. The range of biotechnology options in food and drinks is very wide, and only a few illustrative examples of possible developments are described with, in each case, an opinion on the extent and timing of introduction of these biotechnologies.

A principal factor that will influence the further introduction of biotechnology in foods is the development of the regulatory environment and the associated evolution of

Table 17.   **Food biotechnology developments**

In Commercial Operation	Scientifically Developed – Wider Commercialisation in Next Five Years	Scientifically Feasible – Commercial in 5-10 years
Bioconversion of starch to sweet products	New "functional" foods for specific nutrition needs	Modified food enzymes
High fructose products	Food colours and ingredients by biotechnology	Novel bio-catalysts for food processing
Bioconversion of vegetable oils	Bioreactors for traditional foods, e.g. soya products	Bioconversion technology for unusual food environments
Novel sweeteners	Genetically modified bakers yeast	Genetically modified food bacteria for flavour and quality
Food flavours and enhancers	Plant cell and microalgal culture for high-value ingredients	New biological preservation systems
Fruit juice processing		Rapid "dipstick" tests for all common food contaminants
Amino acids and other special nutrients		
Novel structured foods from fermentation		
Cheese enzymes		
Yeast hybrids		
New biotechnology testing methods		
Lactose-free dairy products		

consumer perception and acceptance of food biotechnology. For this reason, some probable areas of consumer and regulatory concern are considered.

The chapter is primarily concerned with the role of biotechnology in the industrial processing of foods. Inevitably, developments in the processing of foods or in new services associated with the food industry will have some impact on agricultural practice, which will also receive attention.

2. Food Biotechnology – Options

In assessing the impact of biotechnology on food processing and distribution, it must be remembered that there are many traditional and modern biotechnology processes already available for the creation of food products and for improving their quality, nutritional value, or preservation.

Foods are biological by nature and traditional applications include the brewing of beers, the fermenting of wines and the production of distilled liquors. In some cases, such as the raising of bread and the production of cheese, yogurt, salami sausages, pickles, yeast products, soya sauce and fermented soya products, the traditional biotechnology processes have been modernised with industrial process technology and control systems.

There are many other laboratory-based developments for treating foods to improve edibility or preservation or to reduce production costs. Table 17 lists some existing or pending developments for commercial application. It will be observed that the examples are somewhat artificially focused on "off-farm" processing of foods. Further implications for food biotechnology will arise from advances in plant and animal biotechnology, interacting with increased understanding of the nutritional and functional properties of foods. Breeding of plants and animals could in future be linked to highly specific properties of their ultimate food use (see Chapter I, section 6).

Given that a wide variety of food ingredients, food products, food processes or services originate in biotechnology, the question arises as to what conditions will lead to the further application of biotechnology in the food sector. Here there is a need to take account of the complex nature of the food industry. The industry and its supply base are extremely diverse at national and international level. At a global level, food processing ownership ranges from the family level to vast multinational companies. In some cases international companies have grown out of the supplying and processing of primary products such as sugars, meat, milk, starches, edible oils, tea and coffee. Then there are sectors which have developed to supply "traditional" processed foods such as bread, beer or distilled liquors.

There are also major differences between countries and regions in the nature of their food processing and distribution and in the role played by traditional or "modern" food processing. In many countries the food market is further segmented according to developments in the marketing of branded products, including packaged dairy products, meat products, beverages, frozen foods, dried foods and confectionery. There are numerous other industrial groups engaged in the supply of specialist food ingredients such as the amino acids, vitamins, flavours, thickeners and emulsifiers. Biotechnological trends will offer different opportunities and challenges to each of these sectors and to their customers.

A further feature of the food industry is its slow pace of technological change. The major technological steps of pasteurisation, canning and freezing, for example, took decades to spread widely in particular sectors of the industry or from country to country. Partly, this is due to conservatism within the industry. However, the caution of consumers in relation to new technology shown in many food markets has also dictated the pace of technological change. More recently, the wider and more rigorous requirements in food regulations have also impeded the introduction of new technology. This view of slow change in basic technology should not be confused with the rapid and frequent innovations seen in the presentation, format and packing of foods in developed markets. Moreover, advances in other industries find their way fairly rapidly into food processing as, for example, with computerised process control.

Biotechnology may have some special features which allow it to break through in individual sectors. It may provide a reliable supply of an ingredient that is otherwise of intermittent availability, or a more acceptable biological alternative to an existing physical or chemical process. A number of factors now to be considered influence the entry of biotechnology into food sectors:

i) Economic factors. Biotechnology will increasingly offer options of substitution of existing food sources. A well-known example is the supply of sweet products derived from starch, which can substitute for sugars derived from cane sugar or sugar beet. It may also offer cheaper substitutes for scarce or unusual food ingredients such as oils and spices normally obtained from seasonal or wild crops. Inevitably, biotechnology developments will threaten to disturb established trade patterns in certain food materials (see Chapters V and VIII).

ii) Consumer acceptance. Consumer trends in developed countries are thought to be moving steadily towards food products with mild processing and absence of added chemical ingredients. Biotechnology may be able to assist production in line with those trends and can also produce nutritionally improved foods. For example, removal of lactose from milk products by biotechnology is providing a specific benefit for lactose-intolerant consumers. Moreover, development of novel methods for detecting contaminants and new biotechnology preservation systems assist production of safer foods for the consumer. Against this, biotechnology will be perceived by many as new technology, and there is bound to be consumer concern. This may differ depending on the type of biotechnology used, e.g. enzyme processing or genetic engineering. But consumer readiness for the products of food biotechnology will vary from country to country; for example, the Japanese consumer is thought to be well-informed and ready to accept food produced by biotechnology.

iii) Regulatory procedures. Biotechnology in food products, ingredients or processing will raise new regulatory issues. Well-known ones would include the use of genetically modified food bacteria, but others could arise when food components are made in plant cell culture or microalgal culture rather than in normal crop production, while in food processing, the development of enzyme conversion systems may also raise new regulatory questions. These questions will encompass basic confirmation of safety and acceptability and also specific requirements concerning labelling and designation of products. Regulators will need to balance the concerns about new technology against the benefits that may arise from biotechnology in relation to reduction in chemical additives or the improvement of preservation of food products.

iv) New scientific advances. As mentioned earlier, there are many biotechnology steps that can be applied to all types of food. The fact that these are not currently in commercial operation may be due to technical barriers such as yield or purity of processing. Given the recent advances in molecular biology, genetics and enzymology, it is likely that some of these technical hindrances will themselves be removed by the newer understanding of food molecular structure and function. It is unlikely, however, that any one scientific advance will influence all areas of food processing.

v) New bioprocess engineering. In addition to advances in fundamental molecular biology, there are steady technological advances in the design of bioprocess hardware, such as fermenters and control systems. Additionally, some bioprocess chemical engineering may be susceptible to innovative advances such as the use of enzymes as catalysts in non-aqueous solvents. The importance of the latter development is that enzymes may be used for completely novel catalytic actions when applied in organic systems.

vi) Biotechnological services. The foregoing has concerned food ingredients, products or processes. However, biotechnology and particularly some aspects of molecular biology are leading to rapid advances in new methods of laboratory analysis and in the more sensitive or rapid detection of undesirable contaminating micro-organisms, toxins, or chemicals. These developments will provide completely new approaches to testing for quality control and safety assurance. The advances are such that they may revolutionise the approach to control systems in the food industry and provide completely new tools for enforcement authorities in the investigation and enforcement of anti-contamination.

3. Food Biotechnology – Illustrative Examples

In contrast to some other industries, the further exploitation of biotechnology by the food and drink sectors will be much influenced by considerations such as economic value, commercial judgement, consumer acceptance and regulatory and safety considerations. However, the startling advances in molecular biology and all the basic science underlying biotechnology could influence the introduction or rejection of a biotechnology option in particular food sectors. In the following pages examples of food biotechnology options are discussed to indicate where technical advances are likely to influence investment decisions in the coming five to ten years. No one single biotechnology advance will penetrate or revolutionise the whole food and drink industry.

a) Food enzyme technology

Enzymes have many uses in the food and drink industry, including interconversion of food materials, processing of starch, manufacturing of high intensity sweeteners, clarification of drinks, improvement of baking doughs, manufacture of specialised flavours and manufacture of fruit products. Three recent advances likely to find increasing application are enzyme bioreactors, enzyme encapsulation and new enzymes made by protein engineering.

Enzyme processing can be carried out in various ways. In some cases enzyme powder is added to the food substrate, allowed to carry out its catalytic activity, then deactivated by heating. In other processes, enzymes may be fixed onto an insoluble and inert substance, which can be stirred into a liquid food and subsequently removed by centrifuging or filtration. The term immobilised enzyme technology is applied here. Immobilised enzymes can also be packed into sophisticated reactor columns, through which a liquid food stream can pass to allow the enzyme action to take place. The immobilised enzyme is retained on the column. This reactor bed technology permits continuous operation, and is likely to offer cost advantages, together with more control over the required enzymic reaction.

Enzymes can now also be applied in unusual non-water environments, e.g. in organic solvents. This is opening the way to new food process options, such as the design of fats with particular fatty acid composition, or the synthesis of novel peptide flavours. Potential or actual applications of immobilised enzyme technology include: improved beer by continuous brewing, milk processing for cheese, vegetable oil conversions in organic solvents, starch conversion to (much sweeter) high fructose syrup, and lactose removal from whey.

The technology has limitations, however. Immobilised enzyme beds can lose activity on repeated reuse. Costs of immobilised enzyme may limit opportunities, and significant capital investments are required for large-scale continuous reactor systems. These limitations may be offset by improvements in engineering design of the reactors and bed support systems as well as by the development of more powerful enzymes by protein engineering over the coming five or so years (1991).

The reactor systems just described are only suitable for liquid food or drink materials or ingredients. In solid foods, enzymes can be added as surface treatments or by mixing or by injection. However, enzyme activity is difficult to control under these circumstances. Micro-encapsulation of enzymes will offer one important mechanism for controlled application of enzymes in solid food. This has already been achieved with fat-encapsulated enzymes, used to improve the flavour of cheese. Micro-encapsulation techniques for food enzymes are being further refined by the use of liposomes – minute phospholipid spheres made to precise specifications. The liposomes are made to incorporate the desired enzyme. After mixing in the food, the liposome can be ruptured to release the enzyme. Liposomes containing protein or fat-splitting enzymes have been successfully applied to speed the maturation of cheeses and improving their flavour.

b) *Protein engineering of enzymes*

Existing food enzymes can have characteristics that make them less than ideal in a given food processing system. They may have an inappropriate pH response, heat stability or specificity for the job they have to do. Protein engineering will allow the design or alteration of food enzymes at a molecular level. This may involve minor modification, or the design of completely novel enzyme catalysts.

Protein engineering of enzymes requires analysis of protein molecular structure by x-ray crystallography, allied with sophisticated computer graphics. This enables the structure of the enzyme molecule to be defined in three dimensions, and allows the modelling of various possible modifications to the active groups of the enzymes. To bring about the desired modification, site-directed mutagenesis requiring development of a novel gene is used. This novel gene, when inserted in an acceptable micro-organism and expressed by that organism, will provide a recoverable and usable modified enzyme.

A recently reported example of successful engineering of food enzymes was that of the enzyme phospholipase A2. This is an enzyme from a group of interest in the development of improved emulsifiers. Detailed analysis of the positive and negative charges at the end of the molecular helices in the enzyme provided the basic directions for changing heat stability of the enzyme. Specific genes have been restructured for this modified form and inserted into a micro-organism to clone for modified enzymes. Small amounts of enzymes have been recovered and shown to have gained in heat stability, according to the initial prediction. This example emphasizes the fact that protein engineering utilising all the power of modern molecular biology can now be applied to a food enzyme. Clearly this advance will stimulate research and development towards the improvement and modification of other enzymes used in different sectors of the food and drink industry. Protein engineering modifications could be applied to a variety of food processing enzymes, such as those used in starch processing, including glucoamylase, amyloglucosidase, and glucose isomerase. Some applications of protein engineering may come on stream within five years but the biggest impact must be in the five to ten year time horizon.

c) *Biopreservation*

Biotechnology will offer the possibility of biological systems or adapted food micro-organisms that can provide preservation capability in a food, without the use of synthetic chemical preservatives. One interesting example emerging from advanced biotechnology concerns the lactic bacteria. These are widely used in food products, particularly in various dairy products, cheese, cured meat products and also in pickles and in flavour products. Some lactic bacteria have unusual anti-microbial activities deriving from the production of substances such as nisin and pediocin. It now seems likely that these characteristics could be genetically engineered into the food lactic bacteria.

Genetically modified lactic bacteria of food origin may then have potential outside of the traditional fermented foods as preservative agents. It may well be that these lactic bacteria incorporated in a food could provide a fail-safe capacity, whereby if the food was abused by storage at the wrong temperature or for too long, the special lactic bacteria would grow and provide specific protection against undesirable food poisoning bacteria such as the *listeria* or *botulinum*. Molecular genetic research has already revealed the gene responsible for the synthesis on the peptide precursor of nisin. Research is underway to identify the steps which complete the development of the precursor into the anti-microbial nisin molecule. When achieved the identified genes will be transferable into the lactic bacteria normally used for food preparations.

This one example suggests the possibility that a variety of the normal bacteria of food could be genetically modified to provide selected preservation capability in specific foods. They would then provide a route for the intermediate preservation of food or an adjunct to some of the existing chemical and physical preservation technologies. Further examples from developments in recombinant DNA technology can be expected within the next five years.

Introduction of the technology will be influenced by regulatory considerations concerning food bacteria altered by recombinant DNA. Regulatory assessment and consumer acceptance will be eased when – as in the above example – the genetic material being transferred is from within the same species of food bacteria.

d) Food safety rapid contaminant tests

The processing and distribution of food is frequently beset by problems of microbial contamination or the appearance of toxins and other undesirable biological or chemical contaminants. Producers, manufacturers, regulators and enforcement bodies all utilise testing procedures as part of their attempts to identify, control, reduce or remove contaminants dangerous to public health.

Microbiological contaminants include *salmonella, listeria* and *botulinum.* Each of these food poisoning pathogens poses different problems for the laboratory operator for ensuring reliable and rapid detection. In the case of *salmonella* the available testing systems take at least two full days to indicate the likely presence of *salmonella,* while in some cases the testing procedure to confirm the presence or absence of *salmonella* in a sample may last four to six days. For *listeria* this period may be stretched beyond a week to ten days or two weeks. *Clostridium botulinum,* in addition to a traditional culture method, requires confirmation by a toxicity test on mice. The lengthy nature of these tests is not only of concern to the laboratory operator, but poses a serious economic problem for manufacturers requiring efficient screening of materials or products. Complex testing is also a major stumbling block for regulatory authorities attempting to trace sources of contamination.

Advances in biotechnology now open the way to a complete reshaping of the testing procedures for these food pathogens and for other contaminants, through the application of monoclonal antibody detection systems and DNA and RNA probe techniques.

Monoclonal antibodies are derived from special cell lines called hybridomas which can be grown either in culture or in laboratory animals, and in so doing produce an antibody of highly specific reaction characteristics. The antigenic structure of *salmonella* is such that monoclonal antibodies can be constructed that can recognise individual strains of *salmonella.* These antibodies are built into detection systems utilising either fluorescent signalling or a colour reaction (e.g. the widely used ELISA technique) so that the antibody reaction with the specific target antigen on the *salmonella* is very rapidly indicated. Specific monoclonals have been developed to the important *salmonella* organisms including *salmonella enteritidis,* which has been increasing in frequency in cases in the UK and other countries and is an important source of infection in poultry meat, eggs and egg products. These highly specific antibody tests can reduce the testing time for *salmonella* in food to one overnight testing procedure, and may be further refined to detect *salmonella* within a single working day. The shortened test procedures are likely to make major contributions to the raising of safety standards in food supply.

Apart from the monoclonal developments, an important new range of techniques is arising from our understanding of the molecular biology of living systems. It is now possible with important bacteria such as *listeria* to analyse the genetic material in *listeria* cells and then to develop a DNA or RNA probe which will permit rapid detection of the specific nucleic acid signature of the undesired *listeria* contaminant. Much of this is still in the research phase, but the indications are that genetic fingerprinting could revolutionise the early detection of difficult pathogenic organisms in food and greatly increase the sensitivity of tests, so that low numbers of undesirable pathogens could be detected within a few hours of a sample being processed.

As these systems are developed and commercialised in kit form, they will offer new options for food safety control with consequential benefits for manufacturers and also for the regulatory authorities. The developments in basic biological techniques are also providing new options for the detection of other contaminants such as toxins and residues and seem likely to be developed in commercial kit form within the next five years.

e) Novel biotechnology foods

Many biotechnology advances will be focused on the processing or conversion of agricultural produce into marketable food products. Biotechnology is also applied to the direct production of food mass. Large-scale industrial growing of edible mushrooms is an example of the application of modern biotechnology to the production of whole food. However, biotechnology can also offer a means of creating completely novel human foods from lower value materials. An example already achieved at an industrial level is known as myco-protein, pioneered in the UK. This is based on the growth of food grade fungus (*Fusarium* species) on a liquid stream of nutrients. The filamentous nature of the fungus provides an excellent basis for restructuring in an aligned form, to obtain a textured product similar to meat, which can be flavoured and used in recipe products. Under the name of "Quorn" this product is currently being sold in the UK in products which, although like meat, can carry a vegetarian label, and also can be made without animal fats, which is attractive to some consumers.

This development has taken many years, both in technological development and in extensive safety assurance procedures, prior to clearance by regulatory authorities. The history of myco-protein demonstrates that biotechnology can be used to utilise lower-value raw materials, and its history also demonstrates an interesting example of biotechnology providing a unique food product which may meet a particular niche in the demand for speciality premium-priced food products. There are likely to be other examples in the future, but the history of myco-protein clearly indicates the commitment and investment needed to carry through the long development and safety clearance operations. However, there is no evidence that direct biotechnology production of food mass can compete on any large scale in the foreseeable future with food derived from agriculture.

Quite separately, there is a growing trend towards new "functional foods", aimed at specific dietary or health requirements of consumers. Japan is playing a leading role in many of these developments. Examples are nutrient components such as oligosaccharides for low insulin-low calorie response, non-cariogenic disaccharides, "designer" fats, eicosopentaemic acid, and special food fibre ingredients. Biotechnology will play a major role in the design and production of these "functional foods", which will offer special nutritional options for consumers.

f) Plant cell culture

A variety of food ingredients and nutrients can be produced by the technique of plant cell culture. This opens the way to handling plant cells in the same way as yeast or other microbes in fermenting systems. Most attention has been focused on the use of plant cell culture for the production of high-value food ingredients such as aromas, flavours, and colours. The technology is not generally seen to be appropriate for the production of whole foods but offers various options for production processes. These can be based on the growth of free plant cells in liquid cultures, for example in air-stirred fermenters, followed by the separation of the biomass or of the required ingredients from the nutrient liquid. Alternative technologies are under development where, instead of free plant cells in a liquid phase, the roots of plants can be encouraged to grow in a root culture mass in a fermenter vat. In this situation, the root mass is allowed to form a

Table 18. **Plant tissue cultures developed for industrial application**

Product	Species	Company	Country
Shikonin	Lithospermum erythrorhizon	Mitsui	Japan
Berberube	Coptis japonica	Mitsui	Japan
Biomass	Panax ginseng	Nitto Denki	Japan
Peroxidase	Raphanus	Toyobo	Japan
Geranoil	Geranium	Kanebo	Japan
Rosemarinic acid	Coleus blumei	Natterman	Germany
Digoxin	Digitalis lanata	Boehringer Mannheim	Germany

Source: Fontanel and Tabata (1986).

structured mat through which the nutrient material can be circulated and the biological activity carried out.

Plant cell culture developments are not exclusively focused on food ingredients, and a wider range of chemical components can be generated including pharmaceutical and perfumery ingredients. Conceivably, the development of successful food ingredients would be carried out by specialised companies producing a range of high-value low-tonnage products for various industries including the food industry. Table 18 lists a number of plant tissue cultures for secondary metabolites that have been commercialised or are under development for industrial application.

Although the development opportunities have been demonstrated for commercial exploitation of plant cell biotechnology, a number of factors will still influence its successful application. Product yields can be variable and insufficient to make a success-ful commercial operation; scale-up issues in the technology are also critical. The type of fermentation technology can be important: some plant cells are fragile and require sophisticated airlift stirring of the fermentation liquid to ensure desirable yields of the product. Also, there is a high installation cost and high-level management need to ensure successful commercial production.

Further exploitation of plant cell biotechnology will call for greater knowledge of the factors controlling the growth and product excretion by plant cells, and also of how to operate large-scale production systems with cultures that may be unstable. New under-standing of the molecular biology of the control systems in plant cell metabolism will assist in the development of the overall technology, but for some years to come plant cell biotechnology must be seen as a specialised activity for a restricted range of high-value ingredients.

g) Microalgal culture

Growing microalgae produce a wide range of useful compounds, some of which are of importance as human food ingredients or as specialised nutrients. Microalgae have provided a traditional human food in a number of countries (Table 19) and can produce a number of materials suitable for food use, such as colours, vitamins and specialised fats.

Table 19. **Microalgae and some countries where
they are used as food**

Spirulina maxima	Mexico
Spirulina platensis	Chad
Nostoc pruriforme	Peru
Nostoc sphaericum	Peru
Phormiditon tenue	Mexico
Chroococcus turgidus	Mexico
Nostoc commune	Mexico
Lemamea mamillora	India

Source: Pirt (1984).

The large microalgae (seaweeds) are the basis of production of a variety of materials and are a major source of industrial polysaccharides, such as agar, alginates and carrageenin, some of which are used in food manufacture as texture, gelling or stabilising agents.

Microalgal culture has been developed both in a relatively simple pond culture form, and also in more advanced photobioreactor systems. In these industrial systems, the microalgae can be grown in liquid media circulating in transparent sheet or tubes systems that allow the algae to take up available sunlight. Traditional culture of microalgae for human food is well established in some countries, and dried microalgae or their products are also sold widely as a health and vegetarian food in developed countries.

The use of microalgae can lend itself to intermediate and also advanced technology, based on the photosynthetic process. It is likely that advances in bioprocessing engineering will steadily improve the economic basis of microalgal production. The processes will not be confined to high sunlight countries, as they have been shown to be feasible in northern industrial countries. Depending on technological development and regional cost structures, the technology could contribute towards protein production but is more likely to provide the basis for higher-value health food ingredients and fine biochemicals, where market trends favour the biological production of food colours, flavours and specialised ingredients.

Overall, it will be commercial and marketing factors that dictate the expansion of this technology in the next five to ten years, with the likelihood that it will meet certain "niche" segments of the food and food ingredient markets within this time.

4. Consumer Protection

The many possible applications of biotechnology in food processing will lead to different issues in each sector of the industry. Each sector will have to decide on the commercial benefits that biotechnology can offer as against alternative routes for food processing. Food companies will also have to communicate the benefits to consumers by offering improved or cheaper products or alternatives to current processes or ingredients.

Consumer perceptions and judgements will play a critical role in the rate of application of new biotechnology to food. These perceptions and the need for consumer protection will also influence the type of regulations implemented for biotechnology in food.

Biotechnology, by its very nature and diversity, will raise special consumer protection issues which include the following:

i) Safety of a new technology. In many cases biotechnology advances in food processing will be different from all previous processes. The safety assessment of new processes will have to be done for each new case, and there may be relatively few previous data to build on in terms of safety evaluation. Individual countries are already adapting their food assurance procedures to take account of biotechnology developments. These matters are being considered in a separate OECD activity.

ii) Testing of new ingredients. The safety testing is complex and varies depending on the nature of the ingredient. New biological ingredients from biotechnology may appear inherently safe, but doubts may be raised because of lack of information and because proving absolute safety may be impossible.

iii) Ethics of biotechnology. There are differing views on the ethics of genetically altered living systems, and this may influence consumer perceptions. In addition, national and international debate about the safety of release of genetically altered living systems may well slow the application of biotechnology in foods, and particularly the use of altered food micro-organisms.

iv) Biotechnology diversity. There is no one single biotechnology, and the approach to regulation of various different biotechnologies will prove difficult for regulatory bodies. For example, the conceptual and enforcement approach to the control of novel enzyme bioprocessing may be quite different from that required for the assessment and control of genetically engineered food micro-organisms. The conceptual base for covering the range of possible biotechnology processes is so far not well developed.

v) Designation and labelling. Due to the many different ways in which biotechnology products may be brought into the market place, there may be special issues of labelling for biotechnology food products, e.g. where ingredients are made by novel technology such as plant tissue culture of the same species, or where a food ingredient is made in another species by genetic insertion of the required capability in a host such as a fermentation yeast.

There is a great need for widespread communication and debate about the diversity of biotechnology options in food and their implications for safety, consumer protection, consumer perception, regulation and enforcement.

5. Further Reading

BELL, E.R. and E.B. WHITE (1989), "The Potential of Biotechnology for the Production of Flavours and Colours for the Food Industry", *International Industrial Biotechnology*, Vol. 9, pp. 20-26.

DUCASTAING, A. and J. ADRIAN (1990), "Les Enzymes en Technologies Alimentaires", *Sciences des Aliments*, Vol. 10, pp. 231-254.

EDELMAN, J. (1988), "The Development of a New Food", *Journal of Chemical Technology and Biotechnology*, Vol. 43, pp. 279-283.

GERHARTZ, W., ed., (1990), *Enzymes in Industry, Production and Applications*, VCH Vertagsgesellschaft.

FONTANEL, A. and M. TABATA (1986), "Production of secondary metabolites from plant tissue and cell cultures – Present Aspects and Prospects", *Nestlé Research News*, pp. 93-103.

GASSON, M.J. (1990), "Molecular Genetics, Processing and Quality of Foods", Vol. 2, *Food Biotechnology: Avenues to Healthy and Nutritious Products*, 2.255, Elsevier Applied Science.

GASSON, M.J. (1990), "Genetic manipulation of starter cultures", *Dairy Industries International*, Vol. 55, pp. 21-23.

HOFSTRA, H., IN'T. HUIS and J.H.J. VELD (1990), "Bioanalytical Methods: The Application of Monoclonal Antibodies and DNA Hybridisation in Food Microbiology, Processing and Quality of Foods", Vol. 2, *Food Biotechnology: Avenues to Healthy and Nutritious Products*, 2.281, Elsevier Applied Science.

KNORR, D., M.D. BEAUMONT, C.S. CASTER, H. DORNENBURG, B. GROSS, Y. PANDYA and L.G. ROMAGNOLI (1990), "Plant Tissue Culture for the Production of Naturally Derived Food Ingredients", *Food Technology*, June, pp. 71-79.

LAW, B.A. (1990), "Tailoring enzymes for food users", *Food Technology International Europe 1990*, p. 161.

MILLS, E.N.C. and M.R.A. MORGAN (1990), "Using Biotechnology as the Assessment of Food Quality", *Food Technology International Europe*, pp. 227-230.

PAWSEY, R. K. and D.J. COX (1988), "Biotechnology and the Food Industry", Chapter 9, *Molecular Biology and Biotechnology*, eds. J.M. Walker and E.B. Gingold, 2nd ed., Royal Society of Chemistry, London.

PIRT, S.J. (1984), "Algal Photosynthesis: The Aladdin's Cave of Biotechnology", *Chemistry and Industry*, December, pp. 843-849.

SADLER, M. (1990), "Myco-protein – a New Food", *BNF Nutrition Bulletin*, Vol. 15, pp. 180-190.

WASSERMAN, B.P., T.J. MONTVILLE and E.L. KORWEK (1988), "Food Biotechnology", *Food Technology*, January, p. 133.

Chapter IV

NON-FOOD PRODUCTS FROM AGRICULTURE

1. Introduction

Although agriculture is often viewed as targeted solely at food and feed production, it generates a very wide range of other products (Table 20). Agriculture in its broadest sense is a source of biomass, that is, the constituent materials of plants, animals and fish, etc. The production of biomass is not restricted to land; it is also produced from rivers, lakes, seas and oceans.

The actual or potential value of the non-food uses of biomass is often disregarded by policy makers and economists. However, the proportion of agricultural production directed towards non-food uses has already surpassed 20 per cent in value terms in many OECD countries. The global value of the production of forestry and primary forestry industries in 1985 amounted to some $300 billion, around 2.5 per cent of world GDP (Table 21). In the EC, 9 per cent of cereal output finds use in the industrial sector, and world production of cotton, the major natural textile fibre, is greater than that of synthetic textile fibres (Table 21).

Policies to make agriculture more competitive should lay more emphasis on stimulating the non-food sector. This would fit into the long-term trend towards increasing links between agriculture and industry and the emergence of an agroindustrial complex. While agriculture supplies industry, it is also absorbing technological contributions from industry, including services and know-how. Surplus agricultural land in OECD countries is available for non-food production.

In this context the chemical industries play the central role. They develop and sometimes control the chemical and biological technologies which are basic to the processing of all agricultural products. Some non-food products of agriculture have the potential to substitute for synthetic products, if not globally, then at least for specific sectors and products. This may counter the long-term trend for substitution of natural products by synthetic chemical products. Other non-food products such as wood offer a unique technological starting point for industry which has yet to be fully exploited.

This raises a question about the factors that determine competition between natural and synthetic products. These are:

   i) Relative prices of basic raw materials; the main issue being the price of oil compared to that of biomass. Between 1973 and 1986 biomass gradually gained ground against oil, a trend which was reversed following the drop in oil prices after 1986.
   ii) Quality, variability, regularity of supply and safety of basic raw materials.

Table 20. **Some options for the use of agricultural systems**

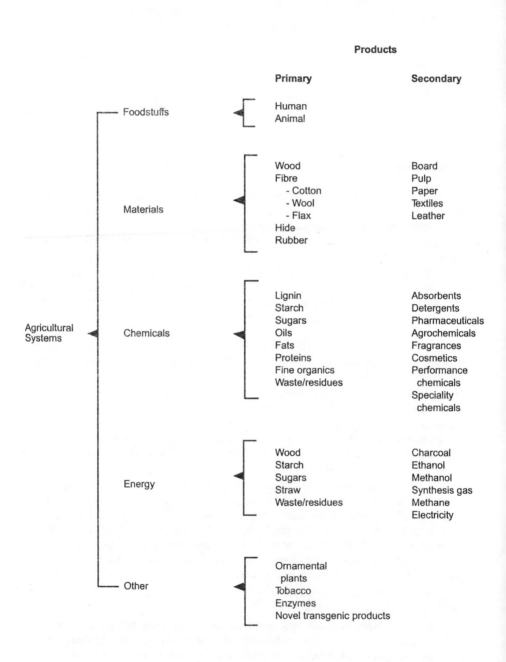

Products

	Primary	Secondary
Foodstuffs	Human Animal	

Materials	Wood
	Fibre
	  - Cotton
	  - Wool
	  - Flax
	Hide
	Rubber
		Board
		Pulp
		Paper
		Textiles
		Leather

Chemicals	Lignin
	Starch
	Sugars
	Oils
	Fats
	Proteins
	Fine organics
	Waste/residues
		Absorbents
		Detergents
		Pharmaceuticals
		Agrochemicals
		Fragrances
		Cosmetics
		Performance
		  chemicals
		Speciality
		  chemicals

Energy	Wood Starch Sugars Straw Waste/residues	Charcoal Ethanol Methanol Synthesis gas Methane Electricity
Other	Ornamental plants Tobacco Enzymes Novel transgenic products	

Agricultural Systems

NB : Wastes and residues can also be upgraded.

Table 21. **Approximate annual world production of some agricultural products**

Sector	Product	Tons	Value in $US billion
Food and Feed	Cereals (including milled rice)	1.8 billion	±250
	Sugar (cane and beet)	120 million	
	Fish	85 million	17
	Crude Starch	1.0 billion	
	Refined Starch	20 million	
Materials	Potential Wood	13 billion	
	Harvested Wood	1.6 billion	
	Fuelwood		100
	Sawwood		60
	Panels		35
	Paper	200 million	110
	Cotton	16 million	14
	Wool	1 million	7
	(Synthetic Textile Fibres)	12 million	
Chemicals	Oils and Fats	70 million	
	Soybean Oil	17 million	8
	Palm Oil	10 million	3
	Sunflower Oil	8 million	4
	Rape Oil	8 million	3
	Starch	2 million	
	Natural Rubber	4 million	4
Other	Cut Flowers and Bulbs		10
	Tobacco	4 million	15

iii) Environmental costs of the entire production cycle; the agricultural-agrochemical cycle is often more polluting of land and water, e.g. via agrochemical residues, than synthetic production.

iv) Relative costs of chemical base material conversion compared to conversion of agricultural products.

v) Premium given by the market to "natural" as compared to synthetic products, and increasing demand for "biodegradability"; these are powerful factors favouring the use of agricultural products by industry.

It should be remembered that the non-renewable fossil feedstocks of oil, gas and coal used by the chemical and energy industries are of biomass origin.

Biomass for the basis of industrial products has various sources such as:

i) Existing natural crops for the production of wood, pulp, paper, polysaccharides, natural fibres, oils and rubber;.

ii) Genetically modified crops for the production of novel materials such as pharmaceutically active proteins.

iii) Waste and by-products of agricultural production and processing, e.g. straw, the residues of grapes which are the basis of tartaric acid production, or rice husks from which furfural is derived.

iv) Agricultural surpluses which can be exploited as feedstocks by the chemical industry for use as a fuel, for the production of ethanol, cellulose, lignin, starch, sugar, etc., although the cost might be prohibitive in some cases (e.g. ethanol).

Non-food uses of agricultural products have considerable potential for underdeveloped and developing countries. In many such countries, especially in Africa, biomass in the form of wood is the major primary energy source. Stocks are being depleted, with grave economic, social and environmental consequences. Biomass and residues of plant and animal origin can be used as a direct (combustion) or indirect source of energy. Examples of the latter are anaerobic fermentation processes to convert sugar from cane and starch from cassava into ethanol, anaerobic fermentation processes to convert agricultural wastes and residues into a gas (biogas) with a high methane content, and direct gasification of biomass coupled with advanced turbine technology for electricity generation.

The development of elite trees for "energy" plantations could ease the pressure on "natural" stocks of vegetation as a source of energy. Moreover, the use of biomass for electricity generation would give less net overall carbon dioxide production than the use of fossil fuels and might be favoured by legislation designed to reduce the rate of carbon dioxide build-up in the atmosphere (to minimise any potential "greenhouse effect"). This technology might find application in developed as well as underdeveloped countries. Major afforestation programmes coupled to the development of advanced biomass power generation technology could reduce the rate of carbon dioxide build-up in the atmosphere at a relatively low cost in terms of global resources.

Biotechnology is very important for the development of non-food uses of agriculture and aquaculture. It can, for example, enable production costs to be decreased, quality and consistency of products to be increased and novel products to be produced.

It should also be remembered that agricultural production and processing leads to the generation of a wide range of wastes, many of which are viewed as major environmental pollutants. Biotechnology has an important role to play in the development of technologies to monitor, minimise, control and treat such wastes.

By definition, the non-food sector is very varied in product and process. The following examples are not meant to be a definitive list of all actual and potential uses of biotechnology in the area, but are illustrative of the ways in which biotechnology can open up new options for producers and consumers.

2. Lignocellulose and Energy

Lignocellulose is a major constituent of plant biomass. Nature's net annual production of plant biomass is estimated at 120 billion tonnes of dry matter on land and around 50 billion tonnes in the world's oceans. Of the biomass produced on land approximately 50 per cent is in the form of lignocellulose.

Commercial forestry and agriculture account for only a small proportion of the planet's biomass production. Total harvested wood amounts to approximately 1.6 billion tonnes per annum (Table 21); potential utilisable wood is estimated at approximately

Table 22. Comparison of scale of energy production as biomass with energy depletion as fossil fuel

Energy Units are given as billions of tons oil equivalent (One tonne oil equivalent = 45.37 Gigajoules)	
Annual sustainable biomass energy production as:	
Sugar	0.05
Starch	0.4
Current Wood Harvest	0.8
Total Plant Biomass on Land	45.0
Annual potential sustainable biomass energy production as:	
Utilisable Wood	6.0
Annual fossil feedstock energy depletion as:	
Total Fossil Energy (Oil, Coal, Gas)	6.6
Total Oil	3.0
Feedstocks for Chemical Production	0.4

13 billion tonnes per annum. There is thus enormous untapped potential both in improved utilisation of existing land biomass resources and in higher plant productivity. In energy equivalent terms, the total renewable resource far exceeds the world's total energy requirements and dwarfs the demand for petrochemical feedstocks (Table 22).

Current large-scale exploitation of biomass as a fuel and chemical feedstock is limited by the low cost of fossil alternatives, the heterogeneous nature of biomass, and its diffuse distribution.

Wood is harvested to provide fuel, structural material and cellulose pulp for paper manufacture. In many developed countries there is a rapidly increasing demand for wood as a structural material. The European Community has a net import of wood to meet demand. The annual global value of the forestry and primary forestry industries is over $300 billion, of which paper production comprises over $110 billion (Table 21). Some cellulose is further modified to provide fibres and thickeners.

It is a sobering thought that over half of the world's population depends on wood for cooking and heating. Developing countries in particular use fuelwood, in some cases getting 90 per cent of their energy from this source. However, an estimated 1.3 billion people cannot get enough fuelwood. Dwindling supplies in rural areas and high prices in urban centres result in a decreased standard of living and in related social, economic, and ecological problems. Deforestation is, in part, the result of the growing energy needs of an increasing population. Ultimately, deforestation often leads to desertification and all its attendant problems.

Biotechnology will have several influences on wood production and processing, ranging from decreased production and processing costs to improved product quality. The agriculture and forest industries are already active in the area of selective breeding, cloning via macropropagation and micropropagation using automation, and genetic modification. Several private and public organisation are involved in the development of fast-growing hardwood species such as pine, eucalyptus and acacias. Eucalyptus is of special promise because in well-managed plantations with early fertilizer application and efficient weed control the economic rotation cycle may be as short as seven years. This is changing the concept of forestry so that foresters are increasingly adopting the practices of the horticulturist and farmer. There is a strong interest in the potential of genetic modification to improve various tree characteristics. The current challenges include the means to transform tree cells and then to regenerate whole plants from this transformed material (see Chapter I). Once these challenges have been met, the production of trees with characteristics such as resistance to herbicides and improved pulping qualities will take the forestry industry into another period of significant change.

The economic effect of small changes can be very significant; for example, on a global basis a 1 per cent decrease in the lignin content of wood for pulp production could give an approximate $1.0 billion saving in annual pulping costs. Progress has been limited not only by the structure and nature of the industry but also by the lack of basic knowledge of tree nutrition, physiology, and genetics. In these areas trees have received much less attention than other economically important agricultural crop plants. However, with better knowledge and management techniques the economics of forestry and the quality of the products should improve. The development of rapid diagnostic systems for disease will aid this.

Improved forestry techniques will be used by producers to gain a competitive edge through reduced production costs and improved product quality. There will be increased international trade in forest products and development of new products and processes based on such materials.

In the longer term it may be possible to develop agricultural and forest "refineries" which will involve whole crop harvesting and use. The harvesting methods are being investigated to reduce waste and to produce high-value by-products from straw, which can be used not only as an animal feed, but also as a feedstock for paper-making, particleboard, and fermentation industries. In current processes of the Kraft type for the production of pulp from wood, the hemicellulose and lignin fractions are mostly lost as low-grade process fuels. The development of cost-effective processes to separate lignocellulose into its component fractions would give a cellulose product, leaving lignin and sugars as by-products for possible upgrading by chemical or biotechnological routes. The current oil price makes competition with base petrochemicals unlikely.

Techniques are available to upgrade the energy value of lignocellulose. The production of charcoal from wood is a traditional example. A newer technology is to densify the wood into pellet form, which can be used as a fuel e.g. for steam-raising. Chemical methods can also be applied. Anaerobic digestion can break down biomass and agricultural wastes into methane and a solid product which can be used as a fertilizer. Small-scale units are already widely known in India and China. Research is continuing, aimed at producing low-cost, highly reliable biomass generators and stoves to improve energy efficiency of local biomass resources. Although small-scale use of biomass is unlikely to make a major contribution to global energy demand, its efficient application could improve the standard of living of two billion people in the developing world, and if coupled with the development of forestry plantations, the pressures on some natural forests might be reduced.

Larger biogas generation units are under development for local electricity generation, and small wood-based cogeneration facilities are an attractive option to provide heat and electricity. In some developed countries, Denmark for example, fiscal regimes allow straw to be exploited as an energy source: 17 per cent of Danish farms have a straw fired boiler, and 20 per cent of the total straw harvested is used for heating.

The rises in oil price in 1973/74 prompted some governments to start programmes for the fermentative conversion of local sugar cane juice into ethanol for use as an automotive fuel. Prospects for these programmes were diminished by the fall in oil prices in 1986. In the longer term, the outlook for such programmes could improve as oil prices rise, and as the range of potential biological feedstocks is extended. By using lignocellulose feedstocks derived from wood, it may be possible to produce ethanol less expensively. Assuming that satisfactory technical solutions can be found for problems associated with pretreatment, cellulose hydrolysis, fermentation of pentoses to ethanol, and the development of lignin-based co-products, ethanol could be produced for as little as $300 a tonne. However a substantial increase in oil price would have to occur before ethanol could compete without any subsidy as an automotive fuel. Ethanol might find better use as an octane booster for gasoline, as it may have less environmental impact than present octane boosters.

The same economic arguments that militate against the use of biomass-derived ethanol as an automotive fuel also militate against its use as a chemical feedstock. Base chemical production from biomass is currently not competitive with the petrochemical industry.

The potential for large-scale exploitation of biomass as an energy source is the subject of feasibility studies on electricity generation from biomass by gasification coupled with advanced gas turbine technology. On the basis of equivalent energy content, electricity in OECD countries costs 3.5 times as much as wholesale automotive fuels, so there is a strong incentive to study the generation of electricity from biomass. The studies indicate that under many scenarios such a process would be competitive with more conventional routes and certainly more competitive than the current nuclear energy options. Such an industry might initially develop on the basis of the 2 billion tonnes of forestry and agricultural residues produced annually. Use of a third of this low-cost feedstock could generate 10 per cent of global energy supply.

In the longer term, biomass could be specifically grown for such a process. As already described for forestry, biotechnology could make a significant contribution towards the reduction of production costs of existing species used for biomass production. Biotechnology could also help the breeding, propagation and culture of novel crops such as elephant grass for biomass production.

Biomass production consumes carbon dioxide. Cropping for power generation diverts biomass from the process of oxidation by decay, which occurs in mature plantations, to oxidation within an energy recovery system: the production of carbon dioxide is virtually unchanged. The use of fossil fuels for power generation produces carbon dioxide. Electricity generation from biomass will therefore produce significantly less net carbon dioxide than its generation from fossil fuels. This could help reduce the rate of carbon dioxide accumulation in the atmosphere and thus help minimise the likelihood of global warming.

The imposition of carbon taxes on fossil fuels would be a strong incentive for the exploitation of biomass as a source of power. A further stimulus would be the introduction of positive carbon tax incentives for biomass production programmes.

3. Starches and other Polysaccharides

Many of the criteria which relate to the applications of biotechnology to lignocellulose also hold for starches, oils and fats.

Approximately 1.0 billion tonnes of starch are produced annually in the form of agricultural products such as corn, potatoes, cereals and rice. More than half is used as food or animal feed after minimal processing. Only about 2 per cent of the starch is processed to pure starch and its derivatives. Of this only about 10 per cent is used in the non-food sector (Table 21). Starch is a feedstock for the production of ethanol by fermentation, as shown by the US programme based on maize. However, in the United States the ethanol is produced at a cost of approximately two to three times that of gasoline, and is economically viable only when subsidised.

The amount of starch processed into industrial products is likely to increase. Such products already find use in applications such as paper-making, textiles, adhesives, and drilling fluids. Recent innovations include grafted superabsorbent polymers (super-slurpers) and cyclodextrins. Carbohydrate derived surfactants (e.g. alkylpolyglycosides) are finding increasing application because of their low environmental impact and toxicity, and their "green" image in the market. There has been a recent upsurge of interest in the blending of starch with synthetic polymers to improve their biodegradability. Environmental pressures on the use of organic solvents, as in paints, will increase the market for water-based systems. This in turn will stimulate the market for aqueous viscosifying agents such as polysaccharides, and for other polysaccharide-derived performance chemicals for application in aqueous systems.

Biotechnology will be of importance in improving the quality of the carbohydrates produced, as an aid to their processing via enzyme technology, and as an aid to their chemical modification. There will be a need for more physico-chemical studies on carbohydrates, especially on polysaccharides and derivatives to enable properties to be matched to potential application. The food industry is already active in this area (Chapter III). Maize, potato and pea plants with different forms of starch (ratio of amylose to amylopectin) due to mutations influencing biosynthetic pathways are already characterised and exploited.

For many plant polysaccharides, technical performance limits application. For example, microbially produced xanthan gum is several times more expensive than plant polysaccharides on a weight basis, but it is cheaper on a performance (viscosifying power) basis. Improved understanding of structure/function relationships might enable modern genetic techniques to improve the performance of plant polysaccharides. The same applies to polysaccharides such as alginate and laminarin from seaweeds and to chitin.

In addition to cellulose and starch, the carbohydrate category includes high-value gelling agents such as agar and alginate from seaweed and chitin, an amino-sugar polymer from the shells of crustaceans, used in modified form as a flocculent.

Environmental concerns over synthetic products could lead to increased exploitation of such natural materials and their derivatives. They could substitute for synthetic products on a performance basis; a whole new class of aqueous-based products could be developed. However, caution is required in equating a "natural" product with low environmental impact; for example, the production of a paper drinking cup consumes more energy and chemicals, and produces more waste, than the production of a polystyrene cup.

4. Hydrocarbons, Oils and Fats

About 70 million tonnes of agriculturally derived oils and fats are produced annually, principally from soybeans, oil palms, rapeseed, and sunflower (Table 21). These find use in foods (80 per cent), soaps and detergents (14 per cent), paints, and a wide range of other applications. Agriculturally derived products are perceived as being more "natural" and "greener" than their petrochemically derived equivalents. For example, fatty acids derived from plants are finding increasing use in the manufacture of detergents, and the cosmetic industry is exploiting wax esters from oil-bearing crops such as jojoba.

Vegetable oils are potential diesel fuel substitutes and extenders. As already described for ethanol from biomass programmes, the major constraint to development is the oil price.

Despite competition from cheaper synthetic rubbers, more than 4 million tonnes of natural rubber (30 per cent of total production) are produced annually for applications ranging from radial tyres to contraceptives. Production of natural rubber is dependent on *hevea brasiliensis*. Biotechnology is playing an increasing role in the process of breeding superior *hevea* varieties, through the development of *in vitro* multiplication methods for superior genotypes. It should also be kept in mind that there are species from some 300 other genera of dicotyledons that synthesize rubber-like polymers of isoprene whose potential has yet to be exploited.

Biotechnology is already playing a role in the improvement of several major vegetable oil producing crops. Oil-seed rape is being transformed into a hybrid crop by the introduction of male sterility systems, either through protoplast fusion or by direct genetic modification. This crop will also be among the first to benefit from the introduction of pest and disease resistance genes. Rapid multiplication of proven elite genetic material, both for agronomic qualities (disease resistance, growth characteristics), and for product quality, could be developed by application of the oil palm micropropagation methods.

Genetic techniques can be used to modify the composition of such plant-derived products, for example to increase the content of short-chain fatty acids in plant oils, or the degree of fatty acid saturation. To develop such modified plants will require not only the development of suitable genetic systems, but also an improved understanding of basic plant physiology and biochemistry, in order to target the genetic modifications to the relevant biochemical steps (see Chapter I). Biotechnology, mainly via enzyme technology, may also improve product extraction and processing.

5. Natural Fibres and Structural Materials

Many natural structural materials are more complex than synthetic materials, and have unique performance characteristics not matched by synthetic materials. For example, on a weight basis many plant fibres have a strength equivalent to glass or carbon fibre. The cost of plant fibres is only 20-30 per cent that of glass, and only 2 per cent that of carbon fibre. Problems of water absorption which detract from performance have recently been overcome, and there is great potential for modified plant fibres in novel composites. Biotechnology can aid the understanding of structure/function relationships in natural structural materials, for example, by the isolation, cloning and amplification of genes responsible for the biosynthesis of natural materials, with a view to the production of individual components in amounts large enough for detailed physical and chemical studies. This will be especially important for natural structural materials produced in very small amounts like the proteins which make up the webs of spiders and give the webs their unique physical and mechanical properties. An improved understanding of natural materials may lead to the improved design of synthetic materials. Biotechnology may be used to genetically modify natural systems to improve structural performance of materials, such as plant fibres. It may also contribute to the processing and modification of natural materials, e.g. by enzyme technology.

Cotton is a good example of a natural fibrous material which has a large market (wood was discussed in section 2 of this chapter). Cotton is the major natural textile fibre, and world production exceeds that of synthetic textile fibres (Table 21). The application of biotechnological techniques to cotton is in its infancy. Cotton accounts for about 10 per cent of the total world's annual consumption of agrochemicals, mainly insecticides. The introduction of insect resistance into cotton (Chapter I) could result in a large reduction in agrochemical use. Longer-term targets are to improve fibre quality, especially length and strength, and to increase the climatic and soil quality "windows" for production. Cotton is an important source of income for many developing countries. This income could be increased by yield improvements, less dependence on chemical inputs and a greater ability to grow on marginal land. Biotechnology could be a powerful tool for the attainment of these goals.

Fibrous proteins such as cotton, wool, hair, and silk are widely used in the textile industry (Table 21), while other non-fibrous proteins find application as glues, emulsifiers, foaming agents, and gelling agents. Leather is essential in the clothing and shoe industries. In the long-term, biotechnological techniques have the potential to improve performance and decrease production costs for many of these products, for example by the development of new vaccines for animal health (Chapter II) or improved enzyme technology for flax processing.

There are many interesting natural structural materials with unique properties that would make them attractive for commercial exploitation. In addition to those already mentioned, there are skin, horn and bone. Some materials may be exploited directly, while others will serve as "models" for fabricating improved synthetic materials.

6. Pharmaceuticals, Agrochemicals and Fragrances

A large share of the annual $100 billion world ethical pharmaceutical market, comprising about 25 per cent of all prescription drugs, is derived directly or indirectly from plants. Steroids make up half the group. Many products of the $20 billion annual agrochemicals sector are also derived from plants. Although plants will continue to provide a source of unique new molecules for both the pharmaceutical and agrochemical industries, the added value is created by R&D in those industries themselves.

Biotechnology is already finding application in the agrochemical industry in two areas additional to those discussed in Chapter I for seeds and microbially based herbicides and pesticides:

i) Molecular biology is a very important technology which enables plant, insect or fungal receptor proteins to be produced in sufficient amounts for chemical and physical characterisation, e.g. by x-ray crystallography. Elucidation of the structure of receptors will facilitate the design of specific active agrochemical molecules that until recently were largely identified by random screening procedures. The "bio-rational" approach should enable more specific, environmentally friendly molecules to be designed, though more fundamental research is needed on plant physiology and biochemistry to discover the important receptors. This is a good example of the enabling role of biotechnology, but it is important to emphasise that advances in technologies such as computer graphics and molecular modelling are also important to the development of this area.

ii) Enzyme technology will be an important tool for the production of new agrochemicals. More than 50 per cent of current agrochemicals on the market are isomeric mixtures. Normally only one isomer is active. Stereospecific enzymes can be used to produce chiral agrochemical products. This will help reduce the input of agrochemicals into the environment. Also, region-specific enzymes can be used to modify molecules chemically where conventional synthetic chemistry is ineffective. However, chemical technologies applicable to this area are also improving, and it is not clear what long-term role enzyme technology will play.

At present most natural fragrances are produced by plants grown mainly in developing countries for markets in developed countries. Biotechnology, in conjunction with existing breeding technologies, may give increased yields. As with flavours (Chapter III), it is now possible to produce some fragrances from plant cell culture under laboratory conditions in developed countries. Because of the poor productivity of such systems, costs are presently high, and despite active research in the area, it is unlikely that such processes will become competitive for some of them, just as in developed countries production in greenhouses is generally not competitive with production in developing countries by traditional methods. It is also unclear whether such products could still be called "natural" and would find acceptance in the market.

Biotechnology is not the only option for aiding production of many of these natural compounds; rapid advances are being made in the area of synthetic organic chemistry. The pyrethroid group of insecticides were originally isolated from plants, but for commercial application they are now entirely manufactured by chemical synthesis. Historically the chemical synthesis of dyestuffs in developed countries caused a major loss of markets for natural dyestuffs produced in developing countries, e.g. indigo production virtually ceased (see also Chapter VIII).

To maximise the prospects for the development of organic molecules from plants, there is a need to maintain a wide germplasm base to maintain genetic diversity, so that we can fully benefit from the very wide range of molecules produced by nature.

7. Ornamental Plants and Tobacco

In the developed countries, the ornamental plant business (which includes cut flowers and lawn grasses) is large and expanding at 12 to 16 per cent per annum. Worldwide turnover for cut flowers and bulbs alone is estimated at $10 billion. The Japanese buy more flowers per person than any other country, closely followed by the Dutch and Italians. Holland leads world exports, with the rose as the top seller in Dutch auctions. The international flower business is attractive to developing countries as it provides export earnings and gives farmers high-value crops to grow. South Africa, Kenya, Zimbabwe, Thailand and Colombia among others, have flourishing export businesses.

Biotechnology has a major influence on the industry. Micropropagation techniques are widely used to multiply disease-free stock and generate identical plantlets. The work is labour intensive, hence when units are large there is a widespread drive to introduce labour-saving automation. Micropropagation technology is being exploited by breeders as well as by plant propagators. Once a new variety has been created and has won official recognition for plant breeders' rights and the royalties that these entail, it is rapidly multiplied by tissue culture for commercial distribution. This can reduce the time for building new stocks from ten years to as little as two and a half years.

The techniques of molecular biology are being used to introduce new traits into given species (see Chapter I). Pure white chrysanthemums could be the first transgenic ornamental plants to reach the market. Whether such constructs will find acceptance in the market is a question of regulatory climate and public perception. Molecular biology techniques will also enable disease resistance to be introduced, which, coupled with advances in diagnostic techniques, will permit better selection of disease-free stock for micropropagation. The development of rapid diagnostic techniques that can be performed *in situ* will help minimise losses during production.

It may seem remarkable that tobacco, which is both one of the major international cash crops (value of $15 billion a year at farmers' prices) and a preferred experimental tool in plant biotechnology, is not a front runner for the economic application of gene technology, although there are some recognized potential applications, e.g. insect and virus resistance. This is mainly due to the almost total absence of an organised seed market for this crop and limited interest. There is no doubt also concern about the regulatory situation and public perception of a genetically modified product. This situation leads to a lack of partners in the seed market for providers of biotechnology such as the small plant biotechnology companies. It illustrates one important point for the application of plant biotechnology in the agrofood sector: namely, that the commercial uptake and development of biotechnology will in general occur only for economically competitive plant varieties that already support significant conventional breeding programmes (see Chapter V).

8. Novel Industrial Products from Transgenic Plants and Animals

Genetic modification techniques can construct plants or animals that produce novel proteins. The production of human serum albumin proteins in potatoes, of mammalian antibodies in plants, and of human blood clotting proteins in animals has already been demonstrated (Chapters I and II). This is a very active area of research, with many small companies in the field. The costs of such products are likely to be much smaller than production costs by traditional routes, or even by bacterial or cultured eukaryotic cell techniques. For example, if antibodies were produced in soybeans and constituted 1 per cent of total protein in soybean meal, a kilogram of antibodies (prior to extraction) could be produced for less than $100. Such developments could open up some new markets and generate a range of new opportunities for agriculture.

Generally speaking, production by agriculture will only be economically attractive for intermediate-value products, such as industrial enzymes. For high-value pharmaceutical proteins, most of the production costs are in downstream processing; and there is little incentive to produce them in transgenic plants or animals unless there is a specific advantage, as in cases where a required structure can only be obtained through the transgenic route.

A complication arises from the strict regulatory requirements associated with pharmaceutical production. These constraints may further militate against the commercial production of pharmaceuticals in transgenic plants and animals.

At the moment, molecular genetic modification and subsequent expression is routinely limited to single genes. There are several technical hurdles to be bridged before multi-gene pathways can be transferred and functionally expressed. Nevertheless, the rate of advance on the genetic front is rapid, though it is difficult to predict the timing of future developments.

9. Miscellaneous Products and Processes

The non-food sector is wide and diverse. Many applications of biotechnology fit into areas which cannot be readily classified in the previous sections and will be briefly discussed here.

Certain enzymes obtained from plants, such as horseradish peroxidase, are currently widely used. The techniques of molecular biology will allow exploitation of plant and animal enzymes, even if they are currently produced at very low concentrations. Production in plants might be increased, or production might be moved into microbes which can be grown under controlled conditions in a bioreactor.

In addition to structural materials (section 5), biological systems produce a wide range of materials with a unique performance. If such materials could be produced in pure form in large amounts they might be exploited commercially, either in their own right or as models for synthetic materials. It is the power of biotechnology, via gene cloning and amplification, to enable large quantities of some such materials to be produced. A good example is the single-chain "antifreeze" peptide which enables the winter flounder to survive in waters as cold as $-2°C$. Such a product might be used to protect California citrus crops, motorways, and aircraft from ice and frost, and act as a cryoprotection agent.

10. Agricultural Waste Treatment and Biosensors

The agrofood industries produce a relatively large amount of varied residues, effluents and wastes which affect the environment. These can be local or diffuse, concentrated or dilute, readily detectable or only detectable with sophisticated technology. They include agrochemical residues, wastes from intensive agriculture, and wastes from processing plants. Many people perceive agriculture to be a major source of environmental pollution, and issues such as the concentration of agrochemical residues in food and water have a high public profile. In OECD countries, regulations related to the area of residues and wastes are increasingly rigorous and comprehensive.

Examples of how biotechnology can help minimise chemical inputs into the agrofood chain have already been given. Biotechnology is also important for the treatment of many effluents and wastes. Many existing waste treatment processes, e.g. the activated sludge process, are biologically based. Biotechnology will play an increasing role in processes such as the following:

i) anaerobic digestion (see section 2) to treat organic wastes and generate biogas as a useful by-product (containing 60 per cent to 70 per cent methane) which can be used as an energy source;

ii) biofiltration technologies which are especially effective in reducing the concentration of malodorous gases from processes such as intensive animal rearing in closed units;

iii) controlled composting to de-water and upgrade many semi-solid wastes, a process preferable to landfill as an option for disposal, especially when there is a market for the composted product;

iv) bioaugmentation, which is the addition of selected microbes to degrade specific wastes, as for example the addition of cellulose-degrading bacteria to wastes from the wood-pulping and paper industries;

v) denitrification processes to remove nitrates from drinking water.

vi) bioremediation technologies to clean up contaminated land and ground water.

These examples indicate the scope of biotechnology for the treatment of residues, wastes and effluents derived from the agrofood industry.

The development of biosensors will aid in the detection and monitoring of residues, effluents, etc. Biosensors are being developed for the detection of agrochemical residues in water and food. They are light, portable, and give a rapid response. They are based on the interaction of the target chemical species with a biological system, e.g. an enzyme, antibody or microbes, under conditions which allow a quantitative physico-chemical response to be measured. Further developments will increase specificity and sensitivity and reduce size and price. Many diagnostic kit and related instrument manufacturers see the development of environmental biosensors and diagnostic systems as their next major market development.

11. Further Reading

ALEXANDRATOS, N., ed., (1988), *World Agriculture: Towards 2000, An FAO Study,* Food and Agriculture Organisation, Rome, Belhaven Press London, UK.

ANDERSON, D. (1991), "Economics of Energy and the Environment – a Discussion of Issues", in *Development Research: The Environmental Challenge,* J.T. Winpenny ed., ODI Publication, January.

CLARK, N. and C. JUMA (1990), "Biotechnology for Sustainable Development", *Policy Options for Developing Countries,* African Centre for Technology Studies Press, African Centre for Technology Studies, Nairobi, Kenya.

ELLIOTT, P. and R. BOOTH (1990), "Sustainable Biomass Energy", *Shell Selected Papers,* Shell Group Public Affairs, Shell Centre, London.

Food and Agriculture Organisation (1983), "Fuelwood Supplies in the Developing Countries", FAO Forestry Paper 42, Rome.

GUNSTONE, F.D. (1990), "Oils and Fats Go to Market", *Chemistry in Britain,* June, pp. 569-571.

HIATT, A. (1990), "Antibodies Produced in Plants", *Nature,* Vol. 344, pp. 469-470.

HOCKING, M.B. (1991), "Paper Versus Polystyrene: A Complex Choice", *Science,* Vol. 251, pp. 504-505.

MACKENZIE, D. (1991), "Energy Answers for North and South", *New Scientist,* Vol. 129, pp. 48-51.

MYERS, N. (1989a), *Deforestation Rates in Tropical Forests and their Climatic Implications,* Friends of Earth Report, ISBN 0 905966 82 2, December, London.

MYERS, N. (1989b), "The Greenhouse Effect: a Tropical Forestry Response", *Biomass,* Vol. 18, pp. 73-78.

National Agricultural Center (1988), "Towards an Agro-Industrial Future", Royal Agricultural Society of England, Monograph Series No. 8, Stoneleigh, Kenilworth, Warwickshire, England.

PERSLEY, G. (1990), *Beyond Mendel's Garden: Biotechnology in the Service of World Agriculture,* The World Bank, CAB International, Wallingford, Oxford, UK.

REXEN, F. and L. MUNCK (1984), *Cereal Crops for Industrial Use in Europe,* Carlsberg Research Laboratories Copenhagen and EC Contract EUR 9617 No. EC1-1001-B7210-83-DK.

WICKENS, G.E., N. HAQ, and P. DAY, eds. (1980), *New Crops for Food and Industry,* Chapman and Hall, London.

World Commission on Environment and Development (1987), "Our Common Future", Oxford University Press, Oxford, UK.

Part Two

INDUSTRIAL AND SOCIO-ECONOMIC ISSUES

Chapter V

STRATEGIES OF LEADING FIRMS IN AGROFOOD

1. Introduction

Previous chapters have concentrated on the acceleration and diversification of new biotechnology applications for the agrofood sector. This scientific and technological potential, however, has to pass the litmus test of economic viability before it becomes a market reality. Nor is it sufficient for a product simply to demonstrate its usefulness or its competitivity. Biotechnology products and processes depend on adoption by leading actors in the agrofood chain, and for this they must be integrated as components of competitive industrial strategies.

Adoption and diffusion of biotechnology in agrofood depend on the global conditions of competitiveness within and even outside the sector. While an adequate analysis of these conditions lies beyond the scope of this chapter, it is important to register that rates of adoption and diffusion depend on the degree to which biotechnology innovations correspond to the major factors underlying current competitiveness. In summary form, this context can be described as one in which there is an excess supply of traditional agricultural products accompanied by a relative stagnation or even per capita decline (both in energetic and monetary terms) in the consumption of traditional foodstuffs in the major industrial countries.

Differential crises in central and eastern Europe, the developing countries of Latin America and the Middle East, and the less developed countries of Africa create a high level of uncertainty as regards the potential for significant geographical expansion of current models of agrofood production and consumption. At the same time, geographical redistribution of these markets depends on complex negotiations to redefine agricultural support policies. Greater levels of trade liberalisation are being stimulated by the financial burden of existing agricultural policies both in the industrialised and the developing countries. Nevertheless, such tendencies are threatened by powerful sectoral lobbies and broader fears as regards socio-political stability.

This pattern of apparent stagnation, however, hides a profound reorganisation in the form and content of food consumption. In the industrialised countries and more generally among middle and high incomes consumers throughout the world, food has now to bear the weight of nutritional, health, pleasure, aesthetic and ecological hopes and anxieties, which involve contradictory/complementary demands ranging from a strengthening of the natural content of foods within the urban-industrial food system to demands for nutritionally engineered foodstuffs.

These demands in their turn have important retroactive impacts within the agrofood chain, both generically in the case of ecology, quality and naturalism, and more specifically in the search for new products. The restructuring of new consumer trends is therefore stimulated by the drive to product differentiation, facilitated by flexible production, transport and communication systems, but also firmly based on a profound redefinition of man's relation to food, itself a component of a broader ecology consciousness.

The dual context which combines both stagnation and the restructuring of the agrofood system provides the background for the different patterns of biotechnology innovation. The analysis of medium-term trends in new biotechnology products, processes and services in this chapter is based on a broad range of in-depth interviews with leading firms in agrofood.

2. Interview Sample

The hypothesis governing our interview sample can be stated as follows: the rate of diffusion of new biotechnologies faces peculiar constraints deriving from the high level of technological and market heterogeneity, combined with an equally high level of interdependence captured in the notion of the "agrofood chain" widely used in the literature.

Patterns of innovation and forms of competitivity therefore vary in accordance with a firm's position in the agrofood system. At the same time, successful introduction of biotechnologies at one point in the agrofood chain depends on their assimilation by other upstream and/or downstream actors. This inter-relatedness of agrofood activities implies that, in addition to competitivity in the immediate markets in which the firms act, the biotechnology innovation must be acceptable within the user markets, particularly that of the final consumer.

One of the main conclusions arising from our interviews and discussed throughout this chapter is that biotechnologies are reinforcing the inter-relatedness of the different economic stages in the agrofood system and that this represents one of the major strategic challenges for innovating firms.

The following objectives therefore guided the interview selection:

i) each of the major markets or economic stages involved in the agrofood system should be represented – seeds, agrochemicals, primary processing, inter-mediates production, and the final food products sector;

ii) a balance between the major economic blocs – Europe, the United States and Japan – should be aimed for, given the specificities of their agroindustrial structure and policy;

iii) some specialist biotechnology firms should also be included, since these have played a determining role in creating the conditions for innovation in biotechnology and its application to the agrofood sector;

iv) the firms interviewed should be among the top ten in their respective sectors.

Table 23 gives a breakdown of the interviews according to the above criteria.

The sample was based on volunteer firms and extensive open-ended interviews and so was subject to limitations of availability and logistics. Given the importance of institutional issues determining innovation patterns, additional interviews were also con-

Table 23. **Leading firms interviewed according to type and location**

Type/Location	Europe	United States	Japan	Total
Seeds	1	2	1	4
Agrochemicals	4	1	0	5
Primary Processing	2	0	0	2
Intermediates	2	0	1	3
Final Foods	3	0	2	5
Specialist Biotech	4	3	0	7
Total	16	6	4	26

ducted within the public sector. A wide range of secondary literature was consulted. This chapter therefore represents a synthesis of the views presented in the interviews, evaluated in the light of the available secondary literature.

3. Summary of the Major Conclusions

Detailed innovation trajectories affecting specific biotechnology processes and products will be discussed in the relevant sections below. Here comments are limited to the general issues relating to agrofood biotechnology as these emerged from the interviews. Five broad conclusions can be drawn:

i) Agrofood biotechnology innovation involves a high level of uncertainty. This derives in the first place from the generic character of the technologies involved, so that competitive options are not limited to questions of efficiency within a clearly defined trajectory. As a result, different technology routes may be mobilised for the same economic objectives. These competitive routes apply across a whole range of potential markets. In the case of biopesticides for example, the *bacillus thuringiensis* may be applied externally with or without genetic modification. On the other hand, plant resistance may be developed through the insertion of genes programmed for the production of specific toxic substances. In the case of hybridisation, male sterility technology confronts self-incompatibility systems. In the area of value-added products, micro-organisms, animals or plants compete as the vehicles for protein production.

Secondly, the radical breakthrough in gene modification has not always been accompanied by appropriate advances in basic scientific knowledge. The gap between gene modification and the expertise necessary to launch biotechnology products and processes may therefore be large. The problems for cost-benefit analysis become more complex when it is understood that expectations have been formed by the technology push and marketing tactics of the speciality gene modification companies.

Finally, uncertainties as regards user response and the regulatory environment are equally powerful inhibiting factors. Similarly, although there may be a considerable homogenisation of the controls over laboratory research, field trials are susceptible to very varied regulation from one country to another. User response on the other hand is intrinsically more ambivalent and can vary considerably for both cultural and other reasons. Optimism on the issue of acceptability was therefore inversely related to the degree of proximity to the final consumer. Also, while moves towards harmonisation are apparent at the level of the general principles of intellectual property, the degree of enforceability is still very unclear (see Chapter VII).

ii) Innovation strategies tend to favour quality processes and products more than cost-cutting applications. The two basic tendencies dominating agrofood markets – the stagnation of traditional outlets and the emergence of new patterns of demand – mean that both cost-cutting and quality-oriented strategies influence biotechnology innovation. Cost-cutting competitive strategies could be seen at work in a number of sectors. In seeds the rapid incorporation of a range of advanced biotechnology procedures is motivated by the need to shorten the time-scale for the production of new varieties. Animal feed is the subject of much research aimed at cheapening conversion costs. The incorporation of enzyme technology to extend maize markets into sweeteners can also be understood along the same lines.

Biotechnology involving gene modification, however, is extremely sensitive from the point of view of acceptability. The issue of visibility is important here and lower visibility may lead to fewer problems, for example in the case of genetically modified biological processing agents. Nevertheless, the difficulties surrounding acceptability would seem to be inducing a shift in innovation strategy away from processes and products which can only be justified in terms of cheaper costs to those where quality benefits can be attributed. Such an innovation strategy is in line with the tendencies towards a restructuring of patterns of food demand and the broader health and ecological concerns discussed above.

iii) Quality product and process innovations and concern with acceptability demand new patterns of co-operation among the different economic actors involved at different points in the agrofood chain. Seed firms claim that farmers are well disposed to the introduction of genetically modified plants that promise lower demand for agrochemicals, since this would lower costs, diminish personal health risks and lessen water pollution levels. Such upstream markets cannot be developed without ensuring acceptance by intermediate and end-users of the resulting agricultural products, where the criteria relate to the health aspects of food consumption, in addition to broader ecological considerations. The successful introduction of biotechnology innovations at any one level of the agrofood system must therefore be negotiated with downstream producers and consumers.

The costs of biotechnology innovation also raise the issue of co-operation at the level of the appropriation of value added among the different agents of the agrofood system. Biotechnology R&D therefore increasingly depends on a revision of the existing patterns of backward and forward linkages, ranging from R&D co-operation and joint licensing to vertical integration.

iv) Successful innovation demands a greater responsiveness to end users whether in the form of other firms or final customers. Reducing uncertainty, the development of quality products and processes that create new markets, new collaborative forms of organisation for sharing risks and benefits, all depend on a greater responsiveness to end-user demand. Where the end-user is industry, new client-customer relations are emerging, but in the case of final consumer demand, the appropriate organisational forms are not so apparent. The issue of biotechnology innovation is leading to an awareness of the need for new forms of co-operation between producer and consumer.

v) Agrofood biotechnology is increasingly dominated, directly or indirectly, by the leading multinationals. Nevertheless, specialised firms are still important for technology transfer and niche markets and will continue to develop as overall markets for biotechnology expand. The diagnostics market is a good example of this latter tendency. The opportunities for small and medium firms will vary, depending on the evolution and structure of the relevant markets.

The above conclusions apply with greater or less force for all the different segments which comprise the agrofood system. The following sections analyse the interview results for each of the major economic sectors involved.

4.   Upstream Markets: Plants

a)   Background

The major focus of attention in agrobiotechnology has undoubtedly been the shifts of the seed market leading to a greater integration with the agrochemicals input sector. This has been the subject of a good deal of analysis whose main conclusions can be summarised as follows. Two waves of purchases into the seed industry occurred in the 1970s. The first was stimulated by the expansion of traditional seed markets in the agricultural boom of the early 1970s, which attracted firms with important distribution synergies. The second wave, involving the major agrochemical groups, was directed to synergies deriving from potential applications of new biotechnologies; it combined a defensive strategic orientation, with a shift away from commodities to speciality chemicals, with closer integration with pharmaceuticals.

b)   Classical breeding vs. plant biotechnology

Much has been made in the literature of the clash between the R&D methodologies of classic plant breeding and plant biotechnology, contrasting the macroscopic approach of the former with the molecular methodology of the latter. There is no doubt that the specific competences of the different actors continue to influence the balance between plant breeding and biotechnology as the relative R&D commitments in Table 24 make clear.

While the various actors have very different strategies with regard to plant biotechnology and traditional breeding, there is wide agreement on the integration of advanced biotechnology techniques as tools for plant breeding. The leading seed firms have all incorporated micropropagation, haploid-diploid, and molecular marker techniques (RFLP) for speeding up traditional breeding programmes and exploiting existing gene pools more efficiently. What is not yet clear is the degree to which new economies of scale are implied in their adoption. This particularly refers to molecular screening techniques, whose full potential was calculated to require very costly and sophisticated computer competence. Specialised biotechnology service firms are emerging in response to demand for molecular genetic screening technology, but the importance of secrecy may give in-house capacity a competitive advantage.

121

Table 24. **R&D expenditure of a selection of seed and agrochemical firms**

Figures in millions of US$ for 1988

Firms	R&D Expenditure	
	Traditional Improvment	Plant Biotechnology
Pioneer	46	7
Sandoz	41	16
Upjohn	24	3
Limagrain	22	5
ICI	21	17
Shell	19	3
KWS	18	5
Dekalb-Pfizer	16	6
Ciba-Geigy	9	17
Monsanto	1	15
Du Pont	0	20
Enimont	0	15
Agrigenetics	6	12
Maribo	7	12
Calgene	1	10
DNAP	0	11

Source: Biofutur, May 1990 (modified and extended on the basis of interview data).

c) *Innovation timetable*

With the more or less simultaneous transformation of corn in 1990 by most of the leading players, the range of crops susceptible to gene modification now includes major cereals and oil field crops, with barley and wheat likely to be included in the near future. Crop transformation on the basis of genetic engineering is therefore no longer a barrier. The question is rather the uses to which gene modification will be put.

In broad terms, the timetable (Table 25) provided by one of the leading firms interviewed for the introduction of new seed products, leaving aside regulatory and acceptance issues, may be taken as a yardstick for the industry's trajectory (see also Chapter I).

While there is general agreement on the technical time scale to market the different innovations indicated in this table, evaluation of the size and importance of these markets

Table 25. **Probable evolution of agrobiotechnology**

1990-1993	Herbicide, pesticide tolerance
1993-1996	Processing improvements
1996-1999	Industrial pharmaceutical production
1999-2003	Environmental tolerance
2003-2006	Direct yield enhancements

varies considerably. We have already seen the widely differing levels of proportional commitment to classical breeding versus plant biotechnology (although Table 25 represents data from 1988, it was corroborated for the current commitment of the firms interviewed in 1990). Higher relative commitment to plant biotechnology is closely correlated with optimism as regards time scale and significance of commercial applications.

Even within what might be considered an optimistic projection, the time scale indicated in the table for direct yield enhancement is in the range of 15 years and over. Within this scenario, world population trends and the slowing down of yield increases from classical breeding would therefore enhance the importance of biotechnology as the provider of new plant breeding tools (see Chapter VIII).

d)   Hybrid technology

Among these tools the most decisive for the creation of new markets would be the competitive technologies for ensuring male sterility as a step to developing new hybrid markets, which would favour companies with large seed investments. While highly specific chemical spraying was still considered to be the most efficient for certain crops (wheat), genetic modification was said to hold the key for future hybrid development.

Two competing genetic packages are now being developed, the male sterility system and the self-incompatibility system. An additional advantage of these systems is that they would eliminate the intensive labour component of current hybrid production. This would avoid the concomitant dangers involved in farming out elite germplasm to low labour cost countries where intellectual property controls are more difficult to enforce than in the industrialised countries.

The male sterility system provides a generic tool involving the same gene and controlling sequences and is therefore applicable to a wide range of crops. Whatever the relative merits of the two systems, there was agreement that these genetic packages would shortly transform the procedures involved in existing hybrid production and open up important new hybrid seed markets. The time scale for application of these innovations now depends on successful negotiation with the actors controlling the relevant markets. This in turn involves complex decisions on the nature of licensing agreements and the feasibility/advisability of moves to vertical integration.

e)   Herbicides

There is a significant level of specialisation in agricultural input markets which influences the market potential of biotechnology applications. Of the three major groups of crop protection inputs, the herbicide market accounts for just under $8 billion, insecticides and fungicides somewhat over $3 billion each. The exact figures are included in Table 26, along with the geographical distribution of the markets according to major economic blocs.

While markets tripled in the 1960s and quadrupled in the 1970s, growth in crop protection markets has slowed down greatly over the last decade. Herbicides account for the lion's share of this market and are product leaders for most of the major agrochemical groups. Scientific and industrial competence in gene modification for herbicide resistance is now widely disseminated, stimulated both by its use as a model system for testing crop transformation and by the potential it offers to expand the market for existing herbicides.

Table 26.

Table 26. **World market for crop protection in 1988**

Millions US$

Zone/Type	Herbicides	Insecticides	Fungicides	Diverse
US, Canada	3 547	1 208	364	346
Europe	2 333	896	1 774	469
Japan, Australia and Southeast Asia	1 444	1 602	1 126	143
Latin America (Brazil, Mexico, Argentina)	640	424	190	31

Source: *Biofutur*, May 1990.

Given the high costs of developing new herbicide molecules, the genetic modification of plants to adapt them to existing market products presents itself as an important alternative.

Major agrochemical and seed firms are actively involved in developing herbicide resistance, either through in-house research or through collaborative agreements and licensing. While most research seems to be directed to the newer, more environmentally friendly herbicides, the fact that all the major agrochemical firms are engaged in biotechnology research to extend herbicide markets has led to the emergence of a critical lobby on the issue (Chapter I). Industry has been put on the defensive, and the specialised biotechnology firms are eager to play down the herbicide label in favour of quality enhancement research.

In spite of their caution on the issue, the agrochemicals firms see no realistic alternative to herbicides. On the other hand, it is recognised that farmers are preoccupied with environmental and health issues involving herbicide use. Farmers, it is claimed, would welcome genetically modified seeds if they cut costs and health risks. If we add to this the strength of the anti-herbicide lobby, it is more likely that marketing strategy will concentrate on those herbicides considered to be less toxic and with a lower application load.

Biotechnology applications would therefore be situated within a more general innovation trajectory towards crop protection involving lower dosage requirements and greater specificity of action. Advances in this direction depend on better knowledge of plant physiology and cell biology. The new genetic techniques therefore become part of a broader strategy that gives pride of place to the biological components of crop protection.

f) Bio-insecticides

The issue of insecticides is more complex, since it involves competing biotechnology routes (plant resistance versus external applications); unlike the herbicide issue, it can represent a substitution rather than a reinforcement of agrochemical markets. Moreover, although all the leading agrochemical firms have important markets in herbicides, it is not the case for insecticides. Indeed, a stimulus for leading firms to invest in insect resistance biotechnology would be the lack of traditional insecticide markets and the opportunity that genetically modified plants offer for entry into these markets. In addition, advanced biotechnology innovations in the field of insect resistance have an important market potential, given the more limited applicability and diminishing efficiency of existing insecticides.

Leading firms committed to existing chemical synthesis of insecticides are sceptical of the large scale substitution potential of biological routes. They see pyretheroids remaining the principal support for an approach increasingly based on management systems and involving the rotation of a range of chemicals on the basis of more precise targeting.

The main alternative at present would be preparations based on the *bacillus thuringiensis* (BT), which accounts for 75 per cent of existing biopesticides but represents less than 0.5 per cent of the global pesticide market. A number of specialised firms have genetically modified BT products either on the market or in the final stages of approval. Patenting is very difficult in this area. Also, traditional BT markets are dominated by only two firms whose aggressive pricing strategy makes market access for product innovation difficult. The pest specificity of biopesticides and the resulting predominance of niche markets increase research and development costs. Nor are biopesticides free from the traditional problems associated with chemicals – pest resistance to BT has already been registered.

An alternative route being developed by firms interviewed involves the insertion of the gene coding for the toxic BT protein into crops susceptible to attack by insects of the lepidoptera species, as in the case of boll weevils in cotton. Here again, commercial application is now in sight, although complex problems of pricing and marketing must first be overcome.

The opportunities for biopesticides increase to the extent that crops are threatened by specific predators that are now increasingly resistant to chemicals or for which there is no chemical solution. The limitation of biopesticides from a technical point of view is the specificity of their action, which is currently largely limited to pests of lepidoptera species. Similarly, it will be less efficient for crops that are threatened by a variety of predators. Economically, BTs are problematic, given the difficulty surrounding patents and the shift to management systems involving crop rotation and a variety of complementary crop protection measures by the leading chemical pesticide suppliers.

Virus resistance is a further important application area for which few agrochemical solutions are available or will be developed. Existing genetic sources of plant protection can be more rapidly incorporated into variety development programmes by the use of rapid screening techniques developed by biotechnology. More recently, genetic engineering has demonstrated the effectiveness of introducing resistance to a range of viruses by recombinant DNA technology. Disease resistance, however, is a complex area, and it is sometimes necessary to combine several independent sources of resistance to provide a durable resistance in the field. Genetic engineering has now dominated the technology of viral resistance for a range of viruses, particularly those affecting vegetable crops.

Resistance to diseases caused by fungi and bacteria is another major field of research that has seen some important achievements. As in virus resistance, the first breakthroughs were in the development of effective diagnostic and screening links. Recent research in plant molecular biology has allowed strategies to be identified by which genetic engineering may provide resistance in major crops to some of the most devastating of these diseases. It will probably take up to 15 years before these developments are translated into new plant varieties (1991).

The crop- and pest-specific nature of biopesticides will tend to create a proliferation of niche markets favouring the development of small specialised biotechnology firms. The difficulty of access to distribution networks controlled by a small number of large firms and the high costs of research make such firms vulnerable to the market leaders. The big markets in cotton or vegetables are already part of the R&D priorities of the leading firms. To the extent that crop protection strategy becomes based on pest management systems, the tendency will be to move away from the single product approach to that of a package of services tailored to specific needs of which biopesticides will be a component.

g) Marketing strategies

How to bring products to market, how to recoup R&D expenditure and how to participate in the value added throughout the different stages of the food chain are the key questions which must be resolved if the costs of biotechnology research are to resist the pressures from venture capital or shareholders. The variety of responses to this question attest to the high level of uncertainty surrounding the potential of biotechnologies but at the same time point to the probable viability of quite different strategies, depending on the specific competences of each firm.

It emerged clearly from the interviews that the rate of return on R&D investment is now becoming critical, since marketed products are the exception and licensing revenues are insufficient when not entirely prejudiced by patent conflicts. Moreover, the traditional seed market is unable to sustain the costs of biotechnology R&D.

Three types of strategies were identified for firms involved in plant biotechnology: the first is that of specialised technology (gene package) suppliers; the second is based on the integration of biotechnology into control over strategic seed markets; the third involves moving downstream to capture the industrial value added which cannot be recouped at the level of seed revenue.

The issues involved are extremely complex and as yet unresolved, with each firm exploring a range of options. The speed of biotechnology diffusion will depend in part on the consolidation of new forms of articulation between the actors at different points in the agrofood chain. The technology supplier option would seem to be the most vulnerable and applies not only to the speciality biotechnology firms but also to major players among the agrochemicals as can be interpreted from levels of commitment to traditional plant breeding shown in Table 24. The logic of this approach derives from the generic character of gene technology (as indicated for instance in the male sterility technology analysed above) as compared with the crop and germplasm specificity of seed involvement. Flexible joint-venture and licensing arrangements would therefore seem preferable to vertical integration. However, the uncertainty of patent legislation, the difficulty of negotiating favourable licensing arrangements and the relatively large number of biotechnology suppliers under pressure for revenue threaten the viability of such a strategy.

A move into seeds is therefore an important avenue to recouping the value added of new biotechnology tools or agronomic traits. In many cases, however, these innovations lead to the substitution rather than the creation of new markets. At the same time, pressure on farm prices limits the price elasticity of new biotechnology products. With the wave of recent purchases, the seed firms have also become extremely expensive, prohibiting the entry of newcomers into the major seed markets. A successful example of

Table 27. **Major core crop markets, 1989**

Million US$

US Input Markets	
Tomato Seed	40
Canola Seed	50
Cotton Seed	50
Alfalfa Seed (Lucerne)	130
Seed Potato Tubers	180
Total	450
US Output Markets	
Processed Tomatoes	400
Fresh Tomatoes	2 500
Edible Oils	3 300
Industrial Oils	500
Alfalfa Feed (Lucerne)	1 800
Total	8 500

such a strategy by one firm interviewed involved the introduction of a seed variety exhibiting important synergies with proprietary crop protection inputs.

All these factors, in addition to the negative image surrounding the close identification of plant biotechnology with agrochemicals, is leading to an increasing concentration on a second generation of products. Besides the more positive image attached to product improvements (nutrition, quality, shelf-life), the industrial value added for crop improvements opens the perspective for attractive returns that would amply justify the high costs of R&D. The relative weight of seed versus agroindustrial markets can be gauged from Table 27 (provided by one interviewee).

Uncertainty still surrounds the acceptability of incorporating genetically modified crops into the food system, an issue which will be discussed when considering the strategies of final food producers. However, while genetically improved tomatoes do not have the appeal of drugs against cancer or AIDS, they no longer carry the stigma associated with pesticides and herbicides. The aim here is to identify biotechnology innovation with the concern for food health and nutrition.

Given the significance of improved crops for industrial value added, the scope for premium pricing arrangements is considerably more flexible than in the case of plant improvements limited to agronomic traits. Nevertheless, the disproportionate economic power of the few buyers of the processing industry, compared to the technology suppliers makes the outcome of such a strategy highly uncertain.

Consequently, there has been a tendency for the core innovating firms to explore various forms of vertical integration. This strategy becomes increasingly viable to the extent that modified crops for industrial use are designed for speciality markets, either as intermediates or final products. Quantities are limited when compared to the commodity seed market and require crops to be subject to separate planting, conditioning, storage and marketing channels. As a result, important high-value niche markets are emerging

127

Table 28.   **Industry research in processing improvement**

Crop	Targets
Tomato	Improved texture Increased solids Enhanced firmness
Potato	Increased solids Reduced browning Uniform starch distribution
Canola	Modified oil composition Increased oil quantity Increased feed quantity
Soybean	Production of pharmaceutical proteins Modified oil composition Increased protein content Increased protein quality Reduced flatulence Reduced anti-nutrients (e.g. trypsin inhibitors)
Cotton	Improved fibre quality
Corn	Increased protein content Increased protein quality Increased oil content Production of speciality starches

that permit patterns of vertical integration and thus greater retention of value added by the innovator firms.

An indication of the scope of "processing improvement" can be gauged by the summary of industry research provided by one firm interviewed (reproduced in Table 28). In the most consistent application of this strategy, one leading firm has developed integrated breeding, planting and processing facilities for each of its major crop facilities. The approach involves a complete restructuring of the competences of the firm involved, but emphasises the increasing fragility of the technology supplier strategy. This tendency reflects the difficulty of isolating new biotechnologies as the basis of commercial strategy.

In this approach, the relevance of new biotechnologies is tied to their competitive advantage in responding to new types of demand. Here the question is not primarily one of comparative costs, although this is sometimes the case (e.g. for increased solid weight of tomatoes), and the returns must be competitive with existing agricultural production. The objective is to occupy new quality markets in terms both of inputs and of final products sought by ecological, health and nutrition concerns. The quality aspects may relate to the ability to control and develop elite germplasm of premium crops that do not require genetic modification. In edible oils, quality would be associated with the relative proportion of mono-unsaturates and poly-unsaturates. Different specialised biotechnology companies are now moving into this market for a variety of crops (canola, sunflower, safflower) and developing in-house, patented hybrids which become part of an integrated agroindustrial operation.

The opportunity here is provided by the emergence of new market demands which break with the mass commodity crop strategy of the leading seed firms and open up niche markets at premium prices. In addition to occupying the high-quality segment of final demand, one company involved identified some 60 niche markets for premium oils. However, while quality objectives based on the genetic modification of seeds may be the medium-term goal, these markets have been won through the development of special hybrids, either using traditional plant breeding or intermediary techniques such as somaclonal variation and protoplast fusion, which allow for a greater exploitation of the existing gene pool. In this context, where the exploitation of nutritional and functional qualities is the objective, new biotechnology tools (haploid diploid techniques, protoplast fusion, gene markers) rather than genetic modification of the final product become the key to competitive advantage.

The ability of science-based biotechnology start-ups to transform themselves into agroindustrial concerns has assumed the potential for a segmentation of bulk commodity grain and oils markets. This requires new patterns of agroindustrial integration to ensure what is known as "identity preservation", involving the separation of the speciality crop during all the post-harvest phases. It also implies new priorities for plant breeding that give a competitive edge over traditional seed firms tied to research programmes based on agronomic traits, with little tradition of downstream integration. The major seed companies are now moving into these markets, and it remains to be seen whether the niche integrated agroindustrial strategy is a viable option for firms still awaiting market confirmation of their scientific investment.

The vegetable oils market has shown itself to be at the centre of current nutritional concerns, but similar research programmes and integrated production strategies extend to carbohydrates and proteins. Here again, transgenic products will have to compete with the development of new varieties with the desired traits drawn from the existing gene pool of the different crops under consideration, using the tools of advanced biotechnology to enhance identification, selection and varietal development.

In the applications time scale presented in Table 25, pharmaceutical products derived from genetically modified plants and animals are placed later than industrial process uses. The latter are indicated as coming on stream in the 1993-96 period, whereas "industrial pharmaceutical production" is projected for 1996-99. However, advances in genetic modification targeting for protein composition may be applied to commodity markets, as in the case of increased lysine content for animal feed crops, or to the cultivation of pharmaceutical proteins. Field tests have already been successfully carried out for molecular farming of the active ingredients of therapeutic drugs. This type of agriculture would have to compete with microbial and animal production routes.

5. Upstream Markets: Animals

a) Background

The 1970s saw two radical challenges to the animal protein economy. First, single cell protein from petrochemical by-products promised a cheap protein alternative that would entirely bypass the meat economy. The leap in petroleum prices coupled with opposition from the grains sector frustrated that initiative, and single cell protein is now reduced to niche animal feed markets. Then meat analogues derived from soybeans were launched as alternative cheap protein sources. These did not gain consumer approval and have now been repositioned as meat complements in the form of isolates and concentrates. Current health preoccupations now broaden the potential of these markets.

Both of the above strategies were stimulated by rigid cost structures, particularly of the beef economy, with its long growth and reproduction cycles and corresponding high feed conversion ratios.

Advanced biotechnologies are now being developed at all levels of the animal food chain – forage crops, animal feed, health, growth and reproduction. These are directed both at overcoming the rigidities which have led to a high cost structure, particularly for red meats, and at the health and quality issues deriving from intensive animal production. The uneven degree of industrialisation of the different protein sources (red meat, poultry, pig meat and fish) provides a variety of options for the incorporation of biotechnologies into industrial strategies.

The shift to intensive breeding in the 1960s and 1970s created major markets for animal health products. Such products are largely spin-offs from human health R&D. It is therefore not surprising that the animal health market is dominated by the leading pharmaceutical companies. The world market, which comprises biologicals, therapeutics and feed additives, is large, reaching a total value of some $9 billion by the end of the 1980s. These markets represent on average only 10 per cent of the leading pharmaceuticals' turnover and often include a wide range of segmented niche markets. Innovation in this sector depends closely on the synergies deriving from research and development in human health.

Paralleling developments in agrochemicals, current patterns of innovation have slowed down under the pressure of regulatory constraints, and competition has increased with the development of low price generic products that are particularly attractive in the context of cost pressures and stagnation in the agricultural sector. The competition from generics has had most impact in feed additives but is now extending to therapeutics. The evolution of the structure of the world animal health market is indicated in Table 29.

Given this competitive and regulatory context, innovation strategies have been directed to biological products, with advanced biotechnologies playing a central role. Vaccines and diagnostics stand out as the main areas. In the case of vaccines, innovations

Table 29. **World animal health market**

Per cent

	1980	1987
Pharmaceuticals	43	47
Feed Additives	45	40
Biologicals	12	13

Source: Adapted from *Animal Pharm 202* (1990).

involve the development of new viral vaccines and delivery systems. Vaccines are also seen as a way of circumventing problems of resistance to antibiotics and parasiticides.

Vaccines are increasingly used in conjunction with diagnostics allowing vaccinated animals to be distinguished from those naturally infected. Here the space for small specialised firms is larger, although major companies are also involved. Pigs have been the leading sector of application followed by cattle, with vaccine delivery systems predominating in the case of poultry. Frost & Sullivan estimate that the combined biotechnology derived vaccines and diagnostics markets will reach $240 million by 1995.

b) Red meat

The red meat sector, which did not benefit from the hybrid technology that revolutionised poultry and pig breeding, is now the subject of major innovations for cheapening and speeding up breeding practices. Through the capacity to freeze bovine semen, artificial insemination has now developed into a world industry and is the basis of 100 million calf births per year. The industry is valued at $2 billion per year and, according to the FAO, artificial insemination has contributed more to animal productivity than improved nutrition, health and husbandry combined.

A second generation of major cattle breeding technology based on embryo transfer is now emerging. High costs have limited the diffusion of this technology. Now *in vitro* fertilization allows for low-cost production. One company interviewed produced over 150 000 bovine embryos in 1990 at a unit cost of less than one-tenth of conventionally produced embryos. Research is directed to freezing technologies to facilitate embryo distribution, to the predetermination of embryo sex, and to the large-scale production of identical embryos.

According to an industrial source, embryo transfer, which involves both male and female genetic stock, will probably replace artificial insemination as the principal method of cattle breeding over the next ten years. The international movement of genetic stock will be greatly facilitated, and the upgrading of herds can be considerably speeded up. Control over sex ratios and the ability to multiply identical embryos will allow for quality improvements and the targeting of production for specific niche markets. Embryo transfer technology has led to the emergence of a new industrial sector with strong representation by small and medium firms.

c) Poultry and fish

Poultry benefited most from the potential of genetic improvements through hybridisation. Rapid industrialisation of the sector occurred on the basis of a combination of genetic, nutritional, and veterinary technologies, and poultry production in the United States now equals that of beef. Since the 1950s the age of chickens brought to market has been more than halved, and the feed ratio has declined by almost as much. Potential for further advances on the basis of these technologies is now limited. However, important advances are occurring in the following areas: efficient vaccine delivery systems based on injections of embryos; peptide injections to improve reproductive physiology; and the preservation and storage of avian semen to facilitate artificial insemination. This potential for growth is being exploited by small specialised firms developing public sector research.

Losses from stress resulting from intensive poultry production has been identified as a critical factor to be addressed for further productivity growth. Experiments have shown that eliminating stress can significantly reduce the age of chickens ready for market. Here the crucial technologies relate to animal health and give a competitive advantage to certain firms, particularly the major agrochemicals, which are able to combine genetics, nutrition and health expertise. As the major agrochemical firms strengthen their position in this sector, the position of traditional agrofood firms in poultry becomes more vulnerable.

Fish represents an important source of protein, and its consumption is being further stimulated by nutritional criteria and the demand for sophisticated foods. The growth in fish farming responds to these tendencies and to the depletion of natural stocks and accounts for 10-15 per cent of landed fish worldwide.

Triploid technology is widely used in certain species to increase growth and quality, and sex control for reproduction is another key area. The markets for veterinary products are in general too small to support the development of specific drugs, but virtually all drugs for higher animals are used where relevant. The synergies involved here favour agrochemicals and pharmaceuticals groups.

Animal feed represents over 50 per cent of total costs, and aflatoxins are a major concern. Important markets therefore exist, both for biotests and for alternatives to traditional fishmeal, providing openings for the by-products of the agrofood processing industry such as yeast by-products from brewing. Additives are another area of application for biotechnology, particularly in the production of vitamin A, which is responsible for fish colouring. Unlike the wild source, vitamin A produced by chemical synthesis is a double isomer and leads to a different colouring. While consumer response to the latter has proved problematical, it is not certain that the biotechnological alternative will be thought natural, even though it is a single isomer that produces the colour of the fish in the wild.

Fish farming is currently limited to a number of important high-value products – carp, shellfish, salmon, trout, prawns, eel (Japan), catfish (US). The Asian market, which continues to grow, accounts for 70 per cent of total production by value. Market saturation now confronts such products as salmon and prawns. Cost pressures have become severe; and while specialised firms play an important role for particular products, there appears to be a tendency towards vertical integration involving major agrofood companies. The diffusion of biotechnologies will depend very much on the marketing strategies of these firms, in an area where the novelty of fish farming makes the issue of ''naturalness'' particularly sensitive.

d) Growth hormone

The most advanced biotechnology products, from the commercial point of view, are in the growth hormone sector involving leading agrochemical firms competing for major global markets. Bovine Somatotropin (bST) is considered to have a market potential of $500 million to $1 billion, with the United States alone accounting for $100 million. The product is currently blocked in the European Community and a number of states in the United States, and market returns from several smaller countries are a long way from compensating the high costs of product development (see also Chapters II and VIII).

The core technology was developed by specialist biotechnology firms in collaboration with the public sector. According to one leading player, bST has no patent protection and is basically a commodity for which the questions of price and service are crucial. Competition depends very much on the relative efficiency of different delivery systems, which range from monthly to daily injections.

Porcine somatotropin (pST) both reduces fat and increases feed efficiency. The same agrochemical and pharmaceutical firms are involved, and the product is now at marketing stage. It is felt that there may be less consumer opposition to its use, and cost-benefit factors favour a speedy uptake by farmers.

Widespread opposition to the use of bST highlights the current sensitivity of consumer and public interest groups to the question of animal welfare. Whether such views will be transitory or not is impossible to say. It is clear that groups involved in final food production and retail are strongly influenced by these trends and are unwilling to be identified with innovations for which consumer response remains ambivalent.

6. Downstream: Intermediates and Primary Processing

While the modification of yeast and enzyme reactions is as old as the earliest cooking and food preservation techniques, their specialised production accompanied the industrialisation of food and washing ingredients. The revolutionary potential of enzyme technology made itself felt with the discovery of the glucose isomerase that permits the production of a sugar equivalent from maize.

a) Enzymes and raw material substitution

Enzymes became equated with a double substitution potential that served to underline the radical promise of biotechnologies. On the one hand, enzymes could be seen as a substitute for conventional chemical extraction processes that use heat across a wide range of bulk products (particularly oils) and involve the degradation of cell walls. Moreover, their application to lignin could be a key component in the economic viability of biomass alternatives for energy production, although this was seen by the interviewees still to be primarily an area of basic research with only long-term commercial prospects.

On the other hand, enzyme technology, as in high fructose corn syrup, was seen to threaten to replace agricultural products, particularly tropical ones. It also represented an alternative industrial strategy to plant biotechnology, which is moving towards increasing fragmentation and specialisation of bulk commodity markets. The first section explained how this process was leading to the development of a speciality cereals and oils market. Enzymes would enhance the options of process technology for blending and transforming a limited number of base materials. The primary processing or agroindustrial sectors would follow this path as they develop technology to create more diversified end-uses for basic staples. Rather than develop speciality oils through plant biotechnology, the approach would favour the enzymatic upgrading of cheap oils as a substitute for high-priced oils, of which cocoa butter is the best known example.

b) Primary processing industries

Although the number of firms interviewed in this sector was very small, the range of tendencies observed point to the radical impact of new technologies and new patterns of consumption on the primary processing industries. These have been directly affected by the interchangeability of starch and saccharose for important sweetener markets. As bulk producers of basic food commodities, they have been most affected by changing patterns of food consumption in the industrialised countries.

The leading actors have countered overproduction and the stagnation of traditional food markets by diversifying the industrial uses of their basic raw materials. While major bulk outlets are still of decisive importance (high fructose corn syrup or HFCS, alcohol) these markets are either reaching maturity or are dependent on government subsidies. An important tendency has therefore been to identify niche markets using process and enzyme technology to transform the basic raw material into specialised intermediates. Modified starches furnish a good example. For the smaller firms, this tendency has reached the point where the product portfolio is determined by client demand.

In the case of the primary processing sector, non-food market outlets have become increasingly important. This is true both for speciality products, particularly in oils and starches, and new bulk markets such as bio-degradable products using a starch base.

In the 1970s and early 1980s, stimulated by the example of isoglucose, radical predictions were made about the potential impact of enzyme technology on the ability to substitute among raw material inputs, particularly those located in developing countries. However, the possibility of large-scale substitution of tropical sourcing through enzymatic transformation or upgrading needs to be qualified. As indicated above, the main thrust to substitution of tropical supplies comes not only from competitive cost pressures but also from nutrition and health factors. This is particularly true in the case of vegetable oils.

In the opinion of one leading firm in this sector, the perspective for a broad-ranging substitution of chemical processes should now be limited to the use of biocatalysis for highly specific reactions. In addition, the price of agricultural raw materials, which constitutes as much as 30 per cent of the final enzyme product, is a further factor limiting competitivity vis-a-vis chemical processes.

c) Market strategy for enzymes

Present industrial strategy in this sector is determined by a number of key elements. In both detergents and food, existing markets can generally be described as mature, so the cost-reducing potential of biological intermediates may be important. This would be true also in the beer industry, where improved yeasts permit the use of lower quality raw materials. However, the growth potential of such markets is limited.

Where major markets emerge in the context of an industrial structure dominated by a few major firms, the tendency seems to be towards an internalisation of demand for intermediates. The leading producers of HFCS are now moving in this direction. The difficulty of patenting in the case of enzymes and yeasts highlights the importance of industrial secrecy and increases the difficulties of co-operation between suppliers and users.

A breakthrough in the traditional enzyme market has been the approval and price competitivity of recombinant chymosin for the coagulation of milk protein in cheese production. This product must compete with rennet from animal and microbial sources and represents a simple substitution of existing sources with no fundamental implications for industrial restructuring.

Faced with the above constraints, the main stimulus to innovation would seem to be the ability to cut costs, and the quality, nutrition, health and ecology factors affecting demand. This is particularly the case in the animal feeds sector. Intensive livestock production is a good example of the combination of cost and new quality considerations. Raw materials represent a high share of total costs in this industry, and its intensive nature has led to high environmental constraints. It is estimated that between 1990 and 1993 new enzyme products will be on the market that simultaneously affect carbohydrate conversion and the release of phosphate, thereby reducing both production costs, and the ecological problems associated with the accumulation of manure in intensive livestock conditions.

d)   Yeasts and flavours

In the case of yeasts, flavour enhancers offer new markets. Yeast technology has advanced beyond hybridisation to genetic modification within the same species. Genetic modification is only one of a range of technologies that include special enzymatic and chemical treatment. For flavour enhancers, the major developments were said to be determined by health criteria, calling in question the use of monosodium glutamate and hydrogenised vegetable oils. Important new markets in snacks and soups for the replacement of meat flavourings are involved and present a challenge to the dominance of leading Japanese firms in the sector.

In the area of flavours, plant cell culture, which is equally applicable to a whole range of secondary metabolites, is also employed. The increasing specialisation of biological intermediates pushes the leading firms beyond the traditional food and detergents markets into pharmaceuticals and the active ingredients of pesticides and herbicides. In the latter case, the preparation of optically active isomers (i.e. removing isomers) is involved, which would considerably lower the chemical load of pesticide and herbicide applications.

Advanced biotechnologies are crucial to the industrial strategies of the leading biological intermediates producers, both for lowering costs in traditional mature markets and for the creation of new specialised markets stimulated by health and ecological considerations. Once bulk suppliers of routine enzymes and yeasts, these firms now see themselves as diversified suppliers of speciality high technology. Commitment to R&D is high, and advanced biotechnology occupies the lion's share. These firms differ from the specialised biotechnology companies in their expertise in the engineering processes of industrial scale-up, which, together with the ability to exploit synergies in widely differing market applications, gives them an important competitive advantage.

Flavours, aromas and colourings have become a key area of biotechnology R&D as opposition to chemical additives has become generalised. In the dairy and bakery industries, research programmes are geared to exploiting the aromatic characteristics of the flora natural to the raw material of these industries, while specialised suppliers are exploiting the potential of secondary metabolites in plants and other living organisms.

e)   Plant cell culture and developing country resource bases

While a reversion to biomass rather than chemical inputs implies a valorisation of agricultural and marine supply bases, especially those located in developing countries, there has been preoccupation with plant cell culture as a substitute for tropical supplies. This perspective was rejected by interviewees from varied industrial and national backgrounds. The level of concern on this issue may of course influence industry's public relations policy, but the range of arguments advanced would suggest that earlier preoccupations were exaggerated (Chapter VIII).

The high costs of plant cell culture technology for the substitution of agricultural raw materials would preclude its use except for high-value, low-volume products, primarily for cosmetic or pharmaceutical purposes. The agricultural supply bases for secondary metabolites have been traditionally based on simple extraction practices, and modernising these supply bases would increase their competitivity vis-a-vis bioreactor substitution. The policy of certain key actors in this sector is to develop direct investments in the strategic agricultural or marine supply bases. Rather than rapid industrial substitution via bioreactor technology, more likely scenarios are either the overexploitation of vulnerable supply bases or the rapid modernisation of these production systems. It was suggested that a more realistic alternative to industrial substitution would be the use of plant cell culture for the development of varieties that could be grown in greenhouse conditions in the industrialised countries. Competitivity would then rest on the economics of greenhouse production.

The secondary metabolites of plants are not themselves patentable, but extraction and manufacturing techniques and applications are; a number of specialised firms have now emerged as suppliers of active plant ingredients. Market strategy here is geared primarily to pharmaceutical or agrochemical applications. However, in the opinion of a leading specialised firm in this area, the major agrochemicals are still committed to existing synthetic approaches.

7. Downstream: Final Food Products

The complexity of the issues involved in the diffusion of biotechnology emerges most clearly in the case of the final food producer sector. This sector traditionally has low levels of internal R&D commitment. Nevertheless, as an earlier OECD study (1979) made clear, it has maintained average levels of industrial productivity through its ability to absorb technology generated in other sectors. This sector of the food industry has been equally affected by processes of restructuring, with a wave of mergers similar to that occurring in other industrial sectors. At first sight, however, product or process innovation would seem to be directed to the control of existing brand products in a move from national or regional to world food markets.

Among the leading firms, R&D commitment has increased as nutrition and health considerations impose new content criteria even on brand products. At the same time, the size of the main firms means that a lower than average level of research commitment does not prevent allocation of significant resources for R&D.

a) Importance of biotechnology

Two basic issues dominate the thinking in this sector as regards biotechnology: the likely consumer response and the potential of biotechnologies for increasing food quality. Public relations are used to influence perceptions of biotechnology, as for instance in the effort to focus attention on the quality of the final products rather than the specific technologies involved in their manufacture. While such criteria may provide the basis for regulatory measures, they cannot guarantee consumer reaction, so caution is observed with respect to producing foods containing genetically engineered ingredients.

While high levels of uncertainty characterise perceptions of consumer acceptability, a large measure of consensus emerged from the interviews that biotechnology must be associated with important gains affecting the availability and nutritional quality of food. There was reluctance to use genetic modification solely for cost reduction. Thus, genetic modification of tomatoes for ketchup would be more vulnerable than the use of similar technology for increasing cassava resistance to endemic disease in Africa. For similar reasons it was argued that biotechnology should not be seen as harmful to the trading potential of the Third World, since the latter represented the strategic markets of the future.

b) *Quality and biotechnology*

It was recognised that the competitive edge in the final food sector had now shifted to issues of quality. While these are difficult to define, they include health, nutrition and ecological aspects. The calculations of one of the world's leading food companies estimate that for 60 per cent or more of consumers in the industrialised countries, quality was the paramount issue. Biotechnologies were identified as important for attaining these goals since they provide powerful tools for tailoring inputs to specific perceived consumer objectives. Similarly, they could provide important competitive advantages to the extent that they offered alternatives to existing chemical processes and inputs.

For specialised firms, seed firms and agrochemical groups, biotechnologies represent a component of new competitive strategies. In the final food sector the resort to biotechnologies will be conditioned by perceptions of consumer response. It is improbable that the key companies in this sector will play an innovative role in the introduction of genetically engineered products. Their response will depend on the evolution of public opinion and perceived consumer response. In the fermentation sector, however, cost pressures and the ability to exploit non-food markets is stimulating investment in biotechnology for the production of bulk and speciality by-products.

c) *Final foods, biotechnology and raw materials*

The competitive potential of biotechnologies for the development of food products and processes corresponding to current images of quality poses a series of difficult decisions for companies in this sector relating to two basic issues: the first derives from the demand for new levels of quality and relates to the degree and the form of involvement in raw material production; the second involves the level of commitment to biotechnology R&D and the institutional form this should take.

The section on plant biotechnology has shown how the issue of quality and the costs of research posed the question of forward alliances and vertical integration. From the standpoint of the final food sector, patterns of backward integration, stimulated by the pressures and potential for quality improvements, have become strategic issues for the leading firms. In all the interviews it was clear that this preoccupation derived from the more general issue of quality requirements. Within this perspective the tools of biotechnology were seen to be decisive for the identification and incorporation of new traits, independently of the issue of gene modification.

For multi-product food firms, the translation of this perception into a competitive industrial strategy is highly complex, and the emerging priorities are specific to the profile of each firm. The heterogeneity of the different agrofood complexes means that the role of the actors and resources is different in each case. Access to elite germplasm, the nature of the seed markets, public sector involvement, all of these factors vary in accordance with particular product groups. The increasingly multi-product character of the leading firms would work against direct involvement in the seed industry. The need for access to germplasm and specific competitive conditions may justify such a move. Irrespective of the institutional forms the alliance may take, the final food sector is increasingly replacing the farmer as the seed industry's privileged partner.

d)   R&D strategy

Among the leading companies a commitment to basic in-house capacity was evident, particularly with regard to strategic product development. Co-operation with specialised biotechnology firms was seen as an option both for specific project development and for access to state-of-the-art technology. At the same time, joint research is pursued with leading actors situated outside the food sector.

The importance for industrial strategy of the uncertainty surrounding consumer behaviour is most evident in the final food sector. The main companies are wary about the acceptability of animal biotechnology. This would also extend to fish, although it is recognised that sexing and triploid technology is already widely applied in the case of certain species, such as trout. It is unlikely that the initiative for genetically modified foods would come from the traditional leading companies in the final foods sector. The issue is less clear in the case of biological agents, particularly when used as substitutes for chemical additives. The decisive importance of biotechnology tools for the competitive development of new quality traits is widely recognised and is the basis for the development of in-house competence, but it must be emphasised that only the largest final food firms have R&D capacity. It remains to be seen to what extent the increasing commitment to in-house R&D will transform patterns of competitivity in the industry.

8.   Leading Firms and Major Economic Blocs: Globalisation

In the interview sample for this report, care was taken to include representation from the three major economic blocs – Europe, the United States and Japan – to reflect their different patterns of agroindustrial development for biotechnology priorities and competitivity. US superiority in the development of plant biotechnology geared to oils and cereal crops can be explained in this perspective. Similarly, European stress on biotechnology related to agrochemicals and Japanese excellence in fermentation products and biotechnology directed to horticulture and secondary metabolites become readily explicable.

The regulatory features of different countries also influence the development of research competence. The strict application of legislation against field trials of genetically engineered products in Japan, for instance, would seem to have led to major advances in cellular biotechnology and the development of new varieties through exploitation of the potential of protoplast fusion. As a result vegetable products from cellular fusion are already marketed in Japan with no negative repercussions.

In examining the position of leading agrofood firms, the dominant characteristic is the globalisation of competitive strategies. Japanese companies avoid the restrictions of domestic legislation through the development of joint ventures or the establishment of subsidiaries in Europe and the United States. Similarly, certain firms in Europe confronted with regulatory restrictions transfer their research programmes to more friendly environments. While front-line research capacity is unlikely to be established in developing countries, the leading firms increasingly have a research and development presence in all of the major industrialised blocs, as the precondition for avoiding specific regulatory limitations and maintaining state-of-the-art competence.

In earlier analyses of global biotechnology, sharp differences were drawn between the USA and Europe with regard to the innovative environment. The paradigm favouring innovation was identified with the presence of a large and diversified specialised science-based biotechnology sector stimulated by well-developed venture capital markets. Within this framework, Europe would be judged negatively on both counts.

The interviews suggested that the in-house competence of the main sectoral players was now the critical factor. This has led to a domestication of biotechnology innovation for the major firms operating on a global scale. At the same time, moves to homogenise the regulatory framework will increasingly provide a more predictable and unified market environment. When added to the tendencies towards globalisation indicated earlier, these factors point to a growing convergence between the competitive innovation strategies of the leading actors.

This conclusion would be particularly true in the case of the leading US and European firms. In the Japanese environment, the lower levels of globalisation of their main companies in the agrofood sector have led to the persistence of important differentiating characteristics. Nevertheless, it must be remembered that the leading firms in agrofood biotechnology are increasingly resorting to collaborative research, industrial joint ventures and the establishment of subsidiaries both in the United States and Europe.

Very low levels of agricultural self-sufficiency and price competitivity have strongly influenced the profile of the post-war Japanese agrofood industry. Horticulture has been the exception, and rice has remained a highly protected staple. As a result, the center of gravity of the food system has been downstream competence in the processing of imported grains, oils and, increasingly, milk and meat products.

In Europe and the United States the technological competence of the processing industry was largely empirical and highly product-specific, having evolved from the scaling-up of traditional artisan techniques. The leading Japanese food industry firms, on the contrary, emerged on the basis of more generic, science-based processing competence dominated by fermentation technology. Exploitation of the potential of fermentation technology, rather than fermenting techniques tied to the processing of a specific product, determined the dynamics of firm growth. This led to the systematic exploitation of micro-organisms for intermediate food use, which transformed Japanese firms into world leaders of commodity intermediates, particularly flavour enhancers and amino acids. It also led to a diversification into major pharmaceuticals markets for antibiotics. Many of the leading firms produce both food and pharmaceuticals, which allows for the continuous exploitation of synergies in R&D.

This very different structure for biotechnology is reinforced by the specific conditions in which it was promoted. Unlike the US and European models where innovation, stimulated by specialised firms, became increasingly appropriated by the leading sectoral actors, biotechnology in Japan was early seen as a strategic option for entry into speciality chemicals. Thus, the major industrial groups all developed a commitment to biotechnology R&D. The fact that many major food firms are formally or informally integrated into the leading conglomerates gives them additional access either to financing or R&D synergies.

The major food firms are therefore simultaneously involved in food and pharmaceutical markets. This is reflected in consumer habits. Per capita consumption of antibiotics in Japan is probably the highest in the world, while the market for physiologically active food and drink provides dynamic niches for innovative firms.

At the level of raw materials, the Japanese agroindustrial sector is adopting an explicit strategy of outsourcing for all bulk commodity inputs. The strategy goes beyond a simple import strategy and extends to the securing of supply bases through credit and infrastructure support schemes, direct investment and joint ventures. This is especially the case for cereals and oils but also for milk and for grapes for wine production. The strategy is combined with the promotion of a high value added domestic agriculture, particularly horticulture for direct consumption and for the production of secondary metabolites. The major seed companies have traditionally concentrated on horticulture, especially vegetables and ornamental crops, where they find a market with diversified demands. Rice and other cereals and grains as well as tuber crops have been largely the domain of the public sector.

New actors are entering the seed sector, led by fermentation firms moving into plant cell culture for the production of secondary metabolites and firms preoccupied by the quality of raw material inputs. This move may lead to joint ventures and foreign investment, or, in the case of milk, to involvement in forage crops and embryo technology for intensive livestock farming. At the same time, the development of new hybrids in rice and the opening of this market to private capital is contributing to deeper involvement in plant biotechnology.

While consumer patterns would seem to be more innovative and open to new biotechnology products, the application of research and release regulations with regard to genetic engineering is said to be stricter in Japan than in a number of European countries or the United States. The major players have countered these limitations by establishing off-shore R&D activities. These restrictions have led to domestic specialisation in plant cellular modification and the successful marketing of vegetable products with new traits achieved through cell fusion technology and the promotion of mutations on the basis of what is called "protoplast breeding".

Globalisation, the potential for collaborative research, and the potential for buying technology on a world scale can compensate for a weaker public sector research base or a stricter regulatory environment. The leading actors are likely to find access to specific front-line biotechnology. As a consequence it may now make less sense to compare formal industrial R&D strengths between the major blocs in this field.

More important perhaps are the factors influencing the rate of mutual penetration of food consumption patterns, which will determine reciprocal patterns of investment in Japan, Europe and the United States. While such tendencies are already apparent, the direct involvement of foreign agrofood firms in Japan is still modest, and the Japanese agrofood giants, with one or two exceptions, have yet to resemble other industrial sectors in the level of their overseas investments.

It appears that in a medium-term perspective the incentives for biotechnology innovation in the Japanese agrofood sector will be influenced by the twin agricultural tendencies of raw material outsourcing and higher value added domestic production. At the same time, the Japanese context provides a special stimulus for biotechnology innovation, as the industrial and research structures reinforce synergies between pharmaceuticals and food.

9. Perspectives for Developing Countries

a) Strategies for raw materials

Sensitivity with regard to consumer response in the final food products sector was also evident in relation to the impact of biotechnologies on developing countries. Some leading companies would go so far as to adopt an explicit policy of not using biotechnology that could harm the trading positions of developing countries. All recognised that fast-growing urban centers in developing countries will represent the strategic medium- and long-term markets.

The use of biotechnology for increasing competitivity between different raw material supply bases and thereby exerting a downward pressure on agricultural prices also raises questions. One of the peculiarities of the agrofood system is the competitive coexistence of processed and non-processed food products. A severe lowering of raw material prices may prejudice the processed food alternatives for which costs are more rigid and agricultural raw material a less important component. This would only apply in the case of specific crops, such as coffee, and to firms simultaneously involved in fresh and processed products derived from the crop in question.

A tendency towards greater promotion of domestic raw materials in food production for developing country markets was also noted. This would partly be a response to demands from developing country governments, but it is also influenced by the impact of changing farm policy on agricultural exports in industrialised countries and the limited capacity of developing countries for financing imports.

The increasingly world-based strategies of leading firms tend to diminish pressures to substitute inputs from the developing countries. Import-substitution biotechnology in the industrialised countries is likely to be stimulated less by multinationals than by regional and national farming and agroindustrial interests.

The interviews with firms involved in the production of food additives, ingredients and biological intermediaries also tend to dedramatise the impact of biotechnologies on developing countries. As has been said, the costs of bioreactor processes exclude basic commodity substitution. While some high value, low-volume ingredients are threatened, the dangers of substitution would seem on aggregate to be more than compensated by the increased demand for natural additives and ingredients. This is leading major companies to increase direct investments in the natural resources of developing countries, which are vulnerable to rapid depletion. On the other hand, the value now being attached to such resources should provide a stimulus to modernising farming and fishing practices in those countries.

b) Biotechnology transfer

The renewed interest of leading firms in the natural resources of developing countries provides an opportunity for the latter to negotiate agreements leading to the internalisation of the skills required for their exploitation. A principal area of concern with regard to biotechnology has been the degree to which scientific and technological competence has become concentrated within the private sector and particularly within the leading multinationals involved in agrofood. It is argued that this development limits the potential for technology transfer and leads to the predominance of R&D agendas which reflect the priorities of industrialised countries (see Chapters I and II).

The interviews point to a more nuanced appreciation of these questions: first, although all leading firms in the agrofood markets have been heavily influenced by the specific domestic market in which they became established, a powerful process of globalisation is affecting their strategic outlooks; second, a key segment of these leading firms sell biotechnology packages; and third, there is a preoccupation with presenting an image of biotechnology as beneficial to developing countries.

Global strategies imply less reliance on domestic resources or markets and the concomitant development of a worldwide scientific and technological capacity. Since this capacity is less tied to local or regional interests, there are fewer constraints on the transfer of technology or on access to products resulting from such technology. A number of leading firms emphasised their competence in frontier technology relating specifically to developing country crops, such as cassava. Other firms indicated their current involvement in research projects directed to problems of developing countries involving collaborative research teams and technology transfer.

The centralisation of frontier R&D programmes, a basic strategy of the leading firms interviewed, does not preclude measures of geographical decentralisation, although only exceptionally has this extended to developing countries. Nevertheless, a policy of training researchers in developing countries for the competitive application of such research in their own countries was evident from a number of the interviews. In many other cases, it was considered that direct purchase of technology by developing countries would be economically and technologically the more viable strategic choice. It was emphasised that a key concern in developing countries was less the transfer of new frontier technology than the diffusion of existing products based on classical breeding and the development of the associated scientific and technological competence.

c) *Institutional preconditions*

The shift of gravity towards the private sector in the generation and diffusion of biotechnology has important institutional implications. All the firms interviewed insisted on the need to develop appropriate legislation for the protection of innovations in this area as a precondition for involvement in research and technology transfer with developing countries, aside from transfers within the global structure of the same firm.

While much of the literature has concentrated on the impact of biotechnologies on the agricultural raw material markets of developing countries, the firms interviewed gave priority to the potential of developing country markets for their products. In a medium- and long-term perspective the newly industrialising countries were seen to be strategic markets, given the saturation of agrofood markets in the industrialised countries.

Three factors limited the rate of current investments and served as a guide to priorities: levels of inflation, the weakness of the local currency, and industrial property rights and their effective application.

High levels of urbanisation, as in the case of Latin America, prevent significant disinvestment even when the above three criteria are all negative. In this case, there is a stagnation of levels of investment. In certain sectors, notably in the case of seeds, the cost and quality aspects of biotechnology research require markets able to sustain premium prices. Even in such a large market as Brazil, the limited demand for quality products for the domestic market makes investment currently unattractive for the leading biotechnology seed firms. On the other hand, strategic growth in the seed market is expected in southeast Asia with Vietnam seen as a potential major supplier of cereals for the region.

10. Conclusions and Policy Implications

One can summarise the general characteristics influencing industrial strategies for the adoption and diffusion of new biotechnology products, processes and services in the agrofood sector as follows:

i) High levels of uncertainty surround the innovation process.

ii) There is a tendency towards development of quality processes and products in preference to cost-cutting applications.

iii) Biotechnology innovation involves new forms of co-operation between economic actors situated at different points in the agrofood system.

iv) Successful innovation demands greater responsiveness to end-users whether they be other firms or the final customer.

v) Agrofood biotechnology is increasingly dominated, directly or indirectly, by the leading multinationals. Specialised firms are still important for technology transfer, and niche markets will continue to develop as the overall market for biotechnology products expands.

vi) The leading multinationals look to worldwide markets for the products of biotechnology innovation and to world sourcing for human and material resources. Such strategies, involving complex patterns of co-operation, allow for the negotiation of both market and institutional restrictions at domestic or regional level.

vii) Substitution tendencies from both plant and processing biotechnologies will continue to influence the competitivity of tropical export crops. The main consequences will depend on the impact of nutritional, health and ecological criteria, as they influence food composition. The questioning of traditional food ingredients may threaten the competitive position of some important tropical cash crops; it may also lead to greater demands on tropical resources for new ingredients and exotic products, thereby accelerating resource depletion or stimulating agricultural modernisation in developing countries. In general, there is a marked dedramatisation of the net impact of substitution processes on tropical crops (Chapter VIII).

While policy recommendations were not directly solicited in the interviews, a number of policy areas can be identified from the nature of the conclusions summarised above:

i) The firms interviewed would give high priority to the definition and harmonisation of the institutional environment affecting biotechnology innovations and their diffusion as a condition for minimising the uncertainty associated with research, field trials, food ingredients and intellectual property.

ii) Stimulus to biotechnology R&D should give a high priority to initiatives that involve new patterns of co-operation between the economic actors at different points in the agrofood system and favour shifts from defensive cost-cutting to quality-oriented products and processes.

143

iii) Biotechnology is in principle a powerful tool for the flexible tailoring of products in response to demand. This potential depends on the ability to integrate consumer interests more closely into the concerns of production. Mechanisms need to be developed which permit advancing from the current defensive perception of consumer interests to a more participatory attitude.

iv) Biotechnologies will play an important part in the necessary restructuring of agroindustrial bases in the developing countries. Access to frontier agrofood technology can no longer be guaranteed within the limits of public sector initiatives. New institutional procedures for technology transfer should be developed which take into account the predominant role of the large private companies and specialist biotechnology firms in the promotion and diffusion of biotechnology innovations.

11. Company Reports Consulted

Agricultural Genetics Company Ltd; Agrigenetics; Amylum; Animal Biotechnology; Calgene; Dalgety; Dekalb; Ferruzzi; Gist-Brocades; ICI; Kyowa Hakko; Limagrain; Monsanto; Nestlé; Pioneer; Plant Genetics; Plantech-Mitsubishi; Rhône-Poulenc; Sanofi; Shell; Snow Brand; Suntory; Unilever.

12. Further Reading

GOODMAN, D., B. SORJ and J. WILKINSON (1987), *From Farming to Biotechnology,* Basil Blackwell, UK.

MAFF (1989), *Biotechnology Research and Development in Agriculture, Fisheries and the Food Industry in Japan,* Tokyo.

MAFF (1989), *Advances of Agricultural Biotechnology,* Tokyo.

OECD (1979), ''Impact of Multinational Enterprises on National Scientific and Technological Capacities: Food Industry'', document, Paris.

OECD (1989), *Biotechnology, Economic and Wider Impacts,* Paris.

PERSLEY, G.J., ed. (1990), *Agricultural Biotechnology, Opportunities for International Development,* CAB International, UK.

US FDA (1988), *Biotechnology and the US Food Industry,* Technomic, USA.

Chapter VI

PUBLIC ACCEPTANCE

1. Introduction

All the chapters in this report dealing with specific agricultural and food biotechnologies address, in one way or another, the issue of public acceptance, particularly of the newer biotechnologies related to genetic engineering, and its importance for the future. However, when reviewing public attitudes and acceptance of agricultural biotechnologies, the analyst is immediately confronted by three major problems. First, the study of public perception of biotechnology has tended not to be very focused, dealing with biotechnology as a whole rather than with specific areas of application like agriculture. If there is a specific focus it has tended to be in the field of human genetic engineering. Apart from the bovine somatotropin issue, there has been little work on public acceptance of agrobiotechnology, although some qualitative research on public attitudes includes substantial analysis of agricultural and food biotechnology.

The second problem is that biotechnology is an umbrella term under which are found an impressive variety of specific technologies and applications. Indeed, almost any meeting or discussion of biotechnology is inevitably prefaced even today by a debate on definitions and the problem of disaggregating the numerous technologies generically known as biotechnology. This is reflected by the rather surprising fact that in EC documents 41 different definitions can be found. The term biotechnology has in fact been around a lot longer than many believe: in 1919 the Hungarian agricultural engineer Karl Ereki defined it as "all lines of work by which products are produced from raw materials with the aid of living organisms". Since then a whole range of definitions have been developed, discussed, revised and argued about. As one historian of biotechnology has put it: "The number of definitions of the word biotechnology has become a matter of embarrassment". If specialists and experts have some difficulty in arriving at appropriate definitions or consensus, then it should not be surprising to discover that the general public is unclear or even confused as to what actually constitutes biotechnology.

The third problem is related to attitudes to science and technology in general. Some studies dealing with public acceptance of science have shown that while a given section of the population may be clearly favourable to science, there may be dissent about specific questions. It would seem, therefore, that very broad questions about acceptance of new technologies (including new biotechnologies) may not provide clues as to how people think about concrete applications. The foregoing considerations have made it extremely difficult to assess the state of public opinion on new biotechnologies as a whole.

2. Quantitative Studies

Recent quantitative social science research might seem to indicate that the public is not well informed and lacks detailed knowledge about new biotechnologies and their specific applications. This conclusion represents an oversimplification of a much more complex situation and may partly be due to the reasons mentioned above, confusion over definitions and the fact that in some surveys the questions posed are too general. There are now a number of major quantitative studies available on attitudes and acceptance of biotechnology in various western countries, including sample surveys in Denmark, Ireland, Spain, and the United States. Perhaps not surprisingly, these surveys contradict one another and do not give a picture of homogeneous attitudes to biotechnology in general or agricultural biotechnology in particular. The diversity of national attitudes and awareness vis-à-vis biotechnology, within the 12 Member States of the European Community and between the formerly separate "east" and "west" areas of Germany, was documented in an extensive Eurobarometer survey conducted in 1991.

A finding common to all the studies with some homogeneity is that the public, insofar as they can, tend to identify biotechnology with genetic engineering and repro-

Table 30. **US study on acceptability of genetic engineering, 1987**

"On a scale of 1 to 10 where 1 is totally unacceptable and 10 is totally acceptable, where would you rank genetic manipulation of ..."

	Average acceptability of genetic manipulation of:			
	Human cells	Animal cells	Bacteria	Plant cells
Total	4.5	5.3	5.6	6.6
Science understanding:				
Very good	5.2	6.1	5.9	7.2
Adequate	4.5	5.3	5.6	6.6
Poor	4.1	4.9	5.4	6.2
Heard about genetic engineering:				
A lot/fair amount	4.9	5.9	6.0	7.2
Relatively little	4.3	5.2	5.4	6.3
Almost nothing	4.3	4.7	5.2	6.0
Effects of genetic engineering:				
Better	5.1	5.8	6.1	6.8
Worse	2.9	4.1	4.3	5.9
Religious:				
Very	4.4	5.2	5.5	6.3
Somewhat	4.5	5.3	5.7	6.8
Not too/not at all	5.1	5.9	5.8	7.2

Source: OTA (1987).

ductive technologies used in human beings. Areas such as deep fermentation ("classical biotechnology") or brewing ("traditional biotechnology") are ignored or not thought to be related to biotechnology at all.

a) The United States

The American study on "Public Perceptions of Biotechnology" (1987) undertaken by the Office of Technology Assessment, which involved a lengthy telephone interview questionnaire, revealed that most respondents expressed generally positive attitudes to science and technology. However, attitudes to biotechnology seemed to indicate a certain ambivalence. Thus, 66 per cent felt that genetic engineering would improve life (92 per cent for solar energy, 51 per cent for nuclear power). While 42 per cent felt that it was "morally wrong" to change the genetic makeup of human cells, only 24 per cent questioned genetic engineering in relation to hybrid plants and animals (see Table 30). Respondents were prepared to accept ecological risks from the release of genetically modified organisms, but opposition was higher for release in the local community. The need for regulation and protection from certain applications of biotechnology that might have negative impacts on environment and health was clearly felt.

b) Europe

In a comparative study undertaken by the European Community as early as 1979, there were significantly different attitudes towards genetic engineering according to country, with 49 per cent in Italy stating that genetic research was "worthwhile" compared with only 13 per cent for Danish respondents (see Table 31).

A Danish study (three separate surveys), carried out between September 1987 and May 1989 in the context of a major parliamentary debate, was based on the OTA questionnaire but focused more on "gene technology". At the aggregate level, ambivalence towards genetic engineering is perhaps the highlight of this Danish survey: "A large majority believe that it is wrong to interfere with the genetic structure of higher organisms. However, an almost equally large percentage feel that questions of gene

Table 31. **EC study on genetic research, 1979**

Percentage replies

	Question: Is genetic research			
	Worthwhile	Of no particular interest	Unacceptable risks	Don't know
European Community	33	19	35	13
Belgium	38	20	22	20
Denmark	13	10	61	16
Germany	22	16	45	17
France	29	22	37	12
Italy	49	20	22	10
Luxembourg	37	31	18	14
Netherlands	36	17	41	6
United Kingdom	29	28	33	10

splicing are best left to the experts... A somewhat smaller percentage believe that it is important not to get behind in technological development. But this is countered by a majority who indicate that they would protest against work on such gene technology being conducted in their neighbourhood. The proposal of an international ban on gene technology receives support from a sizeable minority." The 1991 Eurobarometer survey confirmed that, as in the 1979 survey, Danes clearly had the highest risk perception concerning biotechnology within European Community countries. Of possibly greater interest was the fact that this was combined with "acceptance" levels close to the EC average. The three national surveys indicate that, while risk perception remains high, there has been a continuous trend towards greater acceptance, a trend which it is tempting to attribute to the exceptionally widespread public debate which took place in Denmark over the years 1986 to 1990.

An Irish study undertaken in 1989, using a much simpler and shorter questionnaire than the Danish or American surveys, indicates a very tenuous public awareness of developments in new biotechnologies, for 89 per cent of the respondents were unable to name a single application of biotechnology. This obviously stands opposed to many of the findings of the Danish research, but there is clearly a problem of comparability, given that the questionnaires were different, as was the political context in which the studies were made.

In the case of the Spanish study carried out in 1990, the most notable element is the concern about genetic engineering in humans (49 per cent finding genetic engineering unacceptable in human cells and 64 per cent in human embryos), while genetic engineering in relation to plants, animals and bacteria was seen as positive (for 81, 61 and 78 per cent, respectively). Women were found to be more opposed to genetic engineering on all counts compared with men (15 point difference). Practising catholics were significantly more opposed to genetic engineering than agnostics. The level of educational attainment presented another relevant variable: less education was linked with greater hostility towards biotechnology in general and genetic engineering in human cells in particular. The highly educated seem to have a more positive predisposition to biotechnology at an aggregate level. However, this finding can be misinterpreted or misleading, as will be shown in section 4.

While there seemed to be little concern about genetic engineering in plants and animals, 76 per cent and 72 per cent were against the use of biotechnology for the development of more productive cattle and larger fish. Positive attitudes to biotechnology were found (74 per cent beneficial) although there was clear evidence that many respondents were unsure about what constitutes biotechnology. There has been some debate in Spain, but the mass media has concentrated on developments in reproductive technologies (e.g. choosing the sex of a child and its legal consequences) rather than in agriculture, where the applications in Spain could have far-reaching effects.

In 1991, the Commission of the European Communities supported the already mentioned extensive survey by their Eurobarometer contractors, which involved a 13 question interview of 12 800 adults in the EC countries. The questions focused on awareness of biotechnology in seven areas of application, attitudes towards these various applications, channels of information about new technologies, and trust in various sources. Awareness clearly correlates with GNP, educational level, and individual income. Objective measures of knowledge correlate clearly (and positively) with a favourable attitude towards biotechnology. There was on balance optimism about the effect of new technologies, including biotechnology, on our future way of life; but acceptability of its applications varied widely from medical advances and environmental clean-up (most acceptable) to modifications of farm animals (least acceptable). The world "biotechnology" has in some countries (e.g. Belgium, UK) a more favourable connotation than the words "genetic engineering", an effect which is also more marked in certain application areas (e.g. food processing). Television is clearly the dominant information source, but those naming it as principal source showed a substantially lower objective knowledge than those naming books or specialist publications. Over 50 per cent of Europeans see environmental and consumer groups as trustworthy sources of information on biotechnology; corresponding percentages are 37 per cent for academics; 32 per cent for animal welfare groups; 20 per cent for government sources; 10 per cent for all religious organisations; 6 per cent for industry; and 5 per cent for political organisations. The aim of the survey was to assist efforts to improve the provision of public information about biotechnology, and the Commission has indicated their intention to repeat the survey in 1993. All data have been made publicly available, and professional analysis is continuing.

3. Qualitative Studies

It goes without saying that the socio-economic, cultural and political settings of the above-mentioned countries are very different, and it can be argued that such divergent results are a product of markedly different societies, values and systems of access to information. Another conclusion might be that quantitative studies of public opinion are not always the most appropriate methodology for providing a detailed picture of public attitudes and that other (qualitative) approaches, such as discussion/focus groups, workshops and in-depth interviews, should be used. These quantitative studies have provided analysts with some tantalising clues and have certainly shown the ambivalence of public opinion to new biotechnology applications. The data they provided have been used as the basis for designing much of the most recent qualitative research.

Largely because of the contradictory and sometimes confusing results of earlier quantitative research, qualitative options have recently been used in a four-country study undertaken in Europe under the auspices of the European Foundation for the Improvement of Living and Working Conditions. This has involved research in Britain, France, Germany and Spain and was based on focus/discussion groups of members of the "informed public" (university degree holders, etc.) in each of the four countries, as well as workshops involving representatives of pressure groups and interested parties (pharmaceutical and food processing industries, agricultural interests, medical profession, trade unions, environmental groups, public administrations, etc).

The discussion/focus groups were divided according to sex (there had already been indications from the surveys that responses of men and women were different) and members of the so-called "informed public" were asked to discuss their views on biotechnology. After the initial discussion, they were provided with some basic information about biotechnological applications (bST, biopesticides, herbicide-resistant plants, genetically altered salmon and DNA probes). The group dynamics gave rise to interaction and discussion among the participants and a wide variety of views were expressed. While the groups were politically plural (members of organised lobbies, for or against, were excluded by a preliminary questionnaire), a consensus emerged around a number of issues. In this way the researchers were able to obtain greater insight into acceptance of and anxieties over certain biotechnology applications than they could have from a classical closed questionnaire. It is true the latter has the advantage of reflecting the views of a representative sample of the population, while the discussion group approach cannot pretend to give a precise statistical reflection of the views of the population as a whole at a given time.

In all four countries, participants had heard of biotechnology, mostly in relation to genetic engineering in the medical, reproductive and health care areas. Much less knowledge was evidenced with regard to agricultural and food biotechnologies, with the exception of the German groups, who seemed to have a clearer understanding of the role of biotechnology in these fields. Positive assessments were made of the capacity of biotechnology:

 i) to increase food production, with particular relevance to the Third World;
 ii) to make advances in detection of genetic diseases and introduce new therapeutic techniques;
 iii) to find new techniques to control and treat waste (seen as a possible environmental benefit).

There was some ambivalence about economic benefits and the issue of the profit motive as a principal motivation for biotechnology. Concerns about biotechnology were related to risks and safety. The groups felt that:

 iv) research could get out of hand and adequate institutional controls were lacking;
 v) there is insufficient knowledge of long-term effects of release of genetically altered organisms on the environment;
 vi) certain products, such as herbicide resistant plants, with apparent economic and production benefits might have negative and unpredictable health effects;
 vii) bST in particular was not seen as necessary, because of existing surpluses and because of possible negative effects on health.

Not surprisingly, differences related to sex and country among the groups were found. Women were generally more critical and sceptical than men about biotechnology, especially in relation to genetic engineering. However, the women's groups appeared to trust scientists more than men did where medical and health care was concerned. This applied not only to issues related to biotechnologies in the area of reproduction but also to the food processing industries and agriculture with respect to explicit concerns about safety and public health. Spain was an exception to this pattern.

In Britain and Germany the greatest interest focused on the ethical and medical implications of genetic engineering, but the environmental risks of biotechnology applications were articulated, especially in the German groups, as were concerns about lack of regulatory measures and implementation of existing legislative provisions. In the British case, events in fields not related to biotechnology but to other areas of food production – BSE ("mad cow disease") and salmonella in eggs – seem to have created anxiety among consumers about the will or capacity of the authorities to control activities in the agrofood industry where there are implications for public health and safety. In France, biotechnology was evaluated quite positively (this was reflected in other areas of science and technology, e.g. nuclear power); in Spain the central concern was genetic modification in humans, particularly in the field of reproductive technology. The lack of regulatory measures and state controls were of particular concern in the latter case.

It is possible to speak of positive acceptance of the therapeutic applications of biotechnology (e.g. insulin) but there is concern over possible ethical issues in medical research. At the same time, while response to the role of modern agricultural biotechnology in solving some of the pressing demands of Third World food supply is positive, the public seems to lack confidence in food processing and agricultural biotechnologies as a result of recent but unrelated issues in the United Kingdom and illegal use of growth hormone for fattening cattle in Spain. This is particularly related to the public's doubts about the will of governments to regulate an area where there are possible safety and public health issues at stake; indeed, there even seemed to be a call for international bodies to regulate and supervise developments in biotechnology.

Finally, most groups suggested that there was a lack of clear, understandable and accurate information about biotechnology, and many participants said that they felt unable to comment seriously on biotechnology for this reason. Information and knowledge about biotechnology seems, therefore to be a priority area for the "informed public".

4.   The Context of Public Attitudes

Much of the apparent public concern about biotechnology may be explained by the growing influence of "environmentalism". Major political parties, responding to the growth of the "green vote", are themselves becoming "greener". Environmentalism has become particularly influential in middle to high socio-economic groups and in the most highly educated population groups in OECD countries.

A wide range of environmental issues related to technological changes (including developments in biotechnology) are discernible. In part this is occasioned by uncertainties and lack of confidence in the ability of public administrations to regulate, and of industry to control, activities which can affect the environment and people's health. Waning public confidence is not only related to the effectiveness of regulatory bodies but also to a growing public sense of vulnerability in the face of an apparently increasing number of environmental incidents and risks, some caused by technological innovation.

From the beginnings of modern environmentalism, a central concern has been the effects of the introduction of new agents such as chemicals (e.g. pesticides) into the environment – agents which might have unpredictable consequences on the biosphere or on human health. Already during the early 1960s, some commentators were painting a bleak picture of the planet's environmental future. Such views were shared by all social classes, even those middle and upper status groups which had emerged relatively unscathed from the severe but localised environmental degradation of the early industrial period.

The concern over the impact of applications of new technologies has remained a central theme of environmentalism. Agrobiotechnology has become identified by some groups with environmentally problematic agents such as pesticides, or with completely unrelated food safety and public health issues (salmonella, BSE). Hence, the fact that environmentalism has to some extent been integrated into traditional political discourses could make the introduction of certain new biotechnologies a potential source of conflict. This hypothesis seems to be supported by some qualitative research undertaken on public perceptions of biotechnology.

There are significant differences in public attitudes to biotechnology according to country, sex, educational level, religious practice, etc. The fact that overall the most highly educated groups most readily accept biotechnology does not necessarily mean that they will be receptive to all new applications; indeed, it is precisely in this section of society that the most organised and articulate expression against biotechnology, in the form of environmental movements, is to be found. Thus, public acceptance of new products and applications in agrofood biotechnology will be strongly influenced by the reaction to such developments of the informed members of the public. A key issue in this regard, again confirmed by the 1991 Eurobarometer survey, is the availability to the public of appropriate and understandable information about biotechnology. Furthermore, the opportunity for public participation in assessing the possible value and impact of new technologies needs to be reappraised.

Numerous general or specific proposals have been made during the last years to improve information and communication in this field. Rather than repeat them here, it might be more useful for governments and industry to agree on a number of guiding principles which could help them better to address the issues of public acceptance of agrofood biotechnologies. These principles can be summarised in a number of "points to consider".

5. **Points to Consider** (Formulated by Sub Group I of the OECD Expert Seminar on Biotechnology. Agriculture and Food, 15-17 May 1991, under the chairmanship of Dr. Brian Richards.)

   1. Governments have the responsibility of informing the public about new agrofood biotechnologies which influence choice of products and/or services.

   2. Industry, as an essential partner in this process, has an important role to play in providing the information.

   3. Assessment of the quality and objectivity of the information requires the participation of government, industry, experts and non-specialist interests.

   4. Provision of information to the public has, so far, not been sufficiently effective. New ways and channels need to be developed to communicate information to the public.

   5. Public confidence in regulatory effectiveness needs to be significantly improved.

   6. The public should be made aware of the sustained efforts of scientists and governments to address the potential risks of biotechnology. These efforts have resulted in the safe conduct of a great number of experiments and production in past years and are expected to continue.

   7. The public should, in particular, be informed of different safety assurance approaches developed by national and international bodies. Among the latter, OECD has developed, since 1986 (*Recombinant-DNA Safety Considerations*), principles and criteria for the safe use of genetically engineered organisms in large-scale industrial production and, more recently, in environmental release (Good Developmental Principles).

8. Ethical issues should, as in the past, continue to be identified early and assessed by independent bodies.

9. The end result should be the creation of a favourable climate for balanced judgement on the part of the public with regard to biotechnology.

6.   Further Reading

BORRE, O. (1990), *Public Opinion on Gene Technology in Denmark 1987-89,* Aarhus University, Aarhus.

BUDD, R. (1989), "Janus-faced Biotechnology: A Historical Perspective", in *Tibtech,* Vol. 7, Elsevier, UK.

CANTLEY, M.F. (1987), "Democracy and Biotechnology: Popular Attitudes, Information, Trust and the Public Interest", in *Swiss Biotechnology,* Vol. 5, No. 5, Zurich.

INRA (Europe) S.A. (1991), "Les opinions des Européens concernant la biotechnologie en 1991", report for the Commission of the European Communities on the Eurobarometer survey (English translation in preparation).

Instituto de Estudios Sociales Avanzados (1991), "Biotechnologia y opinion publica en Espana", *CSJC,* Madrid.

Lansdowne Market Research Ltd. (1989), "Biotechnology - Awareness and Attitudes", (Report), Dublin.

OTA (1987), *New Developments in Biotechnology: Public Perceptions of Biotechnology,* US Government Printing Office, Washington DC.

ROBERTS, E. (1989), *The Public and Biotechnology,* The European Foundation for the Improvement of Living and Working Conditions, Dublin.

YOXEN, E. and R. GREEN (1989), "Social and Economic Impact of Biotechnology: Feasibility Study for Future Activities of the Foundation", Mimeo, Foundation for the Improvement of Living and Working Conditions, Dublin.

Chapter VII

PATENT PROTECTION

1. General Considerations

This chapter is limited to the current legal conditions pertaining to patent protection in agrofood biotechnology and to some important recent changes.

Thus, the chapter cannot properly re-examine fundamental principles of patent law and practice, and even less broader questions related to intellectual property rights in general. It is true that patent protection in biotechnology has raised political, ethical and ideological questions which go beyond the narrow confines of patent law, if not law in general. Although some of these questions will be alluded to below, broad and controversial issues, such as that of "farmers' rights" which have been proposed as a reward particularly to Third World farmers, must be addressed at a legal level different from, and higher than, both patent law and plant breeders' rights.

The need for patent protection figures prominently in discussions with industrial companies active in biotechnology. The issue is less dominant than issues of public acceptance and safety, for it belongs to a different political and ethical category and is arcane for the general public concerned about biotechnology. Nevertheless, it has become essential to the industrial development and diffusion of agricultural biotechnology. Patent protection is one of the main conditions for a sufficient financial return on R&D investments, and hence its adequacy does play a role in company decisions on agricultural biotechnology. How crucial this role is in comparison to other factors is difficult to establish in general, but it is noteworthy that insufficient patent protection for some sectors of biotechnology has been an almost permanent industrial complaint and has led to considerable lobbying efforts at national and international levels for nearly 20 years.

Interest in biotechnological patent questions and in the improvement of legal conditions preceded the arrival of substantial numbers of patentable inventions in agricultural biotechnology by many years. Thus, pressure for change was able to make itself felt at political and legal levels before major delays in research and investment or major international distortions had occurred. This does not mean that patenting conditions in biotechnology have so far been without economic consequence. The better conditions afforded by United States law have certainly increased the economic attractiveness of biotechnology investment in the United States.

The present chapter describes recent or planned legal and policy changes, many of which are a direct result of these pressures and proposals from interested parties and legal circles, including those in academia who see in biotechnology the main test case for adapting patent law to an era of rapid scientific and technological progress.

155

Many aspects of patent protection in biotechnology are perfectly satisfactory. The areas of research identified in the earlier chapters give rise to inventions which can be protected by means of process or product patents or by use patents. These types of patent protection, and their strengths and limitations, are familiar from over a century of experience of inventions in other areas of the natural sciences, such as chemistry and especially biological chemistry. This experience can be readily applied to the protection of most inventions in biotechnology with satisfactory results for the inventor. Recognising that inventions involving living material raise special problems, modern patent laws make explicit provision for microbiological processes and products and, by adhering to the various guidelines officially devised for these, protection can often be obtained without undue complication.

2. Agriculture and Patents

a) Existing applications

Although it has not done so in the past, patent law now regards agriculture as broadly on the same footing as any other industry when applying the criterion of industrial applicability or utility in examining the patentability of a given invention. Indeed, European patent law explicitly recognises agriculture as an industry, although it specifically excludes patents for certain innovations related to plants and animals, as will be discussed later in more detail.

Patents can therefore be obtained for a wide range of agricultural and horticultural methods and products, subject to the usual requirements of novelty and inventive step which apply generally to inventions in all fields. Biological agents for the control of agricultural pests, either as pure materials or as formulated compositions, come within the category of product patents. For example, there is an extensive patent literature on the use of *bacillus thuringiensis* strains for this purpose and a number of patents on the use of fungi against weeds (mycoherbicides). Novel techniques of plant micropropagation are protectable as method or process patents. The application of plant cell and tissue culture methods to prepare useful metabolites belongs to the general area of process technology.

In some instances of the foregoing, the procedural rules applicable to microbiological inventions may apply. The principal requirement in this connection derives from the necessity to provide an adequate disclosure of an invention to enable workers of ordinary skill in the art to repeat the process or reproduce the product described. This regulation demands the deposit in officially recognised culture collections of any biological material which is essential for performance of the process, e.g. microbial strains, plant cell cultures, and animal cells, provided these are not already available to others publicly or cannot be fully described and made accessible through the patent description written by the inventor.

The deposit system for "patent cultures" is still a controversial topic, especially as regards the date of availability of the deposit to third parties in relation to the time scale of patent procedure. The dispute has centred mainly on Rule 28 of the European Patent Convention, the prototype for the corresponding rules in the separate national patent laws. This issue is not confined to patents in agricultural biotechnology and requires more detailed treatment than is possible here.

b) The problem areas

The foregoing remarks have presented an encouraging view of the possibilities of patent protection in areas of classical biotechnology where the law generally responds positively. On the negative side of the picture are the exclusions which have been briefly signalled above and which stem from legal policies laid down before the far-reaching impact of agricultural genetic modifications could have been foreseen. These exclusions are to be found in the patent laws of many important countries, but there are major exceptions. This legal diversity is not helpful to research and commercial activity in biotechnology, which is especially international in character, and it is difficult to justify in the present state of the technology.

3. International Practice

a) Legal policies

Differences of patent law from country to country were among the problems addressed in the OECD Report on Biotechnology and Patent Protection (1985). The Report tackled a variety of issues of importance for biotechnology as a whole but focused strongly on the question of patents for plants, a topic on which public debate was already underway. The parallel question of animal patents was not then in the forefront of concern, although it has now become of comparable significance and deserving of urgent attention. The problems addressed by OECD have since been investigated further by other international organisations, notably the World Intellectual Property Organisation (WIPO) and the Commission of the European Communities. WIPO has proposed a series of suggested solutions to these problems (WIPO, 1988) and the European Commission has expressed similar ideas in the form of a draft Directive to Member States (EC, 1988). The WIPO proposals are a basis for discussion on a world scale, including the developing countries, whereas the EC Directive has the more immediate practical objective of improving and harmonising the national laws of EC Member states in preparation for the single market.

In both the WIPO and EC documents the remit is of wide scope covering inventions involving biological material of all kinds. Within this broad canvas, however, the problems of patent protection in agricultural biotechnology stand out as being of central importance. In particular, the proposed EC Directive is resolutely aimed at raising European standards of legal protection as far as possible to the levels enjoyed under the more favourable laws of the USA and Japan, especially in regard to patents for genetically modified plants and animals.

b) US law

Recent developments in the patent case law of the USA have led to a maximalist solution which has its origin in the US Supreme Court dictum that patents can be allowed for "anything under the sun that is made by man" (Chakrabarty, 1980). This pointed the way to confirmation that utility patents (the normal type of patent for industrially applicable inventions) can be obtained for plants, as in the case of a patent for a tryptophan overproducing maize obtained through tissue culture (Hibberd, 1985). Under US law the same plant can, in some circumstances, be protected by a patent granted by the US Patent Office as well as by a Certificate of Variety Protection issued by the Department of

Agriculture under other legislation. To obtain a patent the plant must embody an invention, whereas for the certificate the ordinary tests for new varieties apply.

The Hibberd case and another one concerning a method of producing a polyploid oyster (which did not succeed for other reasons) were followed by the general statement of the US Patent and Trademark Office approving the patentability of "non-naturally occurring non-human multicellular living organisms, including animals" (Commissioner D.J. Quigg, 1987), and the consequential issuance of the onco-mouse patent (Harvard, 1988).

The current position in the United States appears to be stable as regards utility patents for novel types of plants, so long as patentability criteria are met. On animal patents, however, attempts are being pursued by farming, other special interest groups and by groups opposed to animal patents for ethical reasons, to challenge the policy of granting such patents and to secure legislative exemptions which will severely detract from their commercial value.

c) Japanese law

Although the question of patents for plants and animals has been controversial, it is admitted that the Japanese patent law does not preclude such patents and some indeed have been issued. Japan also has a Seeds and Seedlings law, and plants may be protected under either this law or the patent law or both in appropriate cases.

d) European law

In contrast to the US and Japanese laws, the patent laws of Europe contain a number of specific exclusions. Among these is the exclusion of patents for "plant and animal varieties and essentially biological processes for the production of plants and animals".

The prototype of this exclusion is Article 53(b) of the European Patent Convention (EPC), which is the normative patent statute for most of western Europe, and it is upon this particular provision that controversy has been sustained. Until recently the debate has centred on the plant aspect of Article 53(b), and it is therefore more convenient to deal with this separately.

4. Legal Protection for Plants in Europe

a) Plant breeders' rights

The origin and purpose of the prohibition of patents for plant varieties stems from the earlier policy of traditional plant breeders and the agricultural industry to operate a system of legal protection of narrower scope and therefore less strong than that of patents. Hence arose a parallel system of legal protection for plant varieties (known as plant breeders' rights or plant variety rights) under national laws and the International Convention for the Protection of new Varieties of Plants (UPOV). The UPOV Convention allowed member states to grant either patents or plant variety rights for the same entities but prohibited the grant of both types of protection simultaneously (now referred to as cumulative protection).

The reason that this form of protection is weaker than that of patents is that the rights are, to a great extent, limited to the commercialisation of the reproductive material (seed or vegetative) of the specific variety. Also, the farmers and breeders were allowed certain freedoms. Under the "farmers' privilege" the farmer is allowed to take seed obtained from the first plant crop (produced from seed bought from the seed merchant) and to resow this derived seed on his own farm. He can repeat this use of farm-saved seed as often as he wishes, subject to any controls based on other regulations that may apply. Under the "breeders' privilege", also called the "research exemption", other breeders are allowed to take the protected variety and to breed from it a new variety which they are quite free to commercialise. It should be noted, however, that these freedoms have since been reviewed because among plant breeding circles it has been recognised for some time that the law is in need of reform.

In relation to plant varieties, the legislators of the European Patent Convention (EPC) went a stage further than UPOV and enshrined the deliberate exclusion of patents for plant varieties in the patent law itself, in Article 53(b) noted above. In the absence of a clear definition of the term "variety", which is not a recognised botanical scientific group (taxon), and the even greater uncertainty over the term "essentially biological", the meaning and scope of the above exclusion could only be determined by judicial authority or by further clarifying statute law. To add to the difficulty, Article 53(b) also states that the exclusion does not apply to "microbiological processes and the products thereof".

Judicial interpretation came first from a case involving a patent claim to seeds and other plant propagating material treated with a chemical compound protective against certain herbicides (Ciba-Geigy, 1984). In holding that this claim was not forbidden by Article 53(b), the decisive authoritative tribunal of the EPC judicial system (Technical Board of Appeal) addressed the wider issue and in effect concluded that the scope of the exclusion corresponded with the areas of protection obtainable under the plant variety rights system. The latter form of protection is limited at the level of particular varieties shown to be distinct from other known varieties, primarily in phenotypical characters, and also uniform and stable. This interpretation was confirmed in a later case involving a method of producing hybrid varieties (Lubrizol Genetics, 1988) in which it was also held that the claimed method was not "essentially biological" because the defined sequence of steps neither occurred in nature nor corresponded to classical breeders' processes.

For the purpose of agricultural biotechnology this particular question is perhaps significant only to the extent that it is at all relevant to plants produced by the methods of biotechnology as distinct from those of classical plant breeding. Genetic modification by recombinant methods, for example, will produce new "types" of plant material which are not yet developed to the stage of the variety but which form the parental material from which varieties will eventually be bred. Under the EPC therefore it seems that the outcome will depend entirely on the wording of the patent claim and whether it is to be construed as directed to a variety, as such, or to some other level of classification which cannot be refused under this particular Article.

b)   The response of the plant breeders

Plant breeders have come to accept that the protection available under plant variety rights has itself to be strengthened by extending its scope and foreclosing avenues of too easy avoidance of these rights. Moreover, with the advent of plant biotechnology, which plant breeders are now themselves beginning to use, it can be seen that patent protection has its attractions and may be the only feasible type of protection for the technical developments which are taking place.

UPOV has recently been revised (March 1991). Former Article 2 prohibiting double protection (patent and plant breeder's right) has been removed from the Convention, although Member States will be free to retain it in their national law. However, against the wishes of the majority of observers representing industrial groups, a definition of the term "variety" has been adopted (Article 1) which is not limited to varieties that qualify for protection under UPOV. The object and function of such a broadened definition within the context of this legislation is difficult to ascertain. A retroactive application of this change to widen the scope of the exclusion of plant varieties from patent protection is likely to be strongly resisted by both official and interested patent circles. Its possible legal effect must therefore be a matter of conjecture until tested.

Other amendments of UPOV in the recent revision considerably modify the application of the breeders' and farmers' privileges. These changes will be widely welcomed by the innovative elements of the agricultural industry. The strengthened rights under UPOV will however still fall short of those afforded by patents and will not cover process technology. It is also arguable that the two forms of protection are fundamentally distinct in their legal nature so that they do not present a legal interface problem. Instead, the two forms of protection can be seen as complementary rather than as conflicting.

c) The views of patent circles

There is widespread recognition throughout patent circles that the achievement of crop and animal improvement by means of genetic engineering must be reflected by appropriate forms of legal protection which satisfy the needs of the innovating laboratory in academic as well as industrial research. An increasing number of plant breeders and official spokesmen of UPOV recognise this also and do not question the role of patent protection in this respect.

The view has been held for some years in industrial and professional patent circles that a plant variety should be patentable provided it meets the criteria of patent law. It has also been urged that cumulative protection should be allowed provided the criteria under each system are fulfilled. These ideas are also gaining ground in official patent circles, a development which is encouraging to those who have held their ground throughout this time. It should be noted that the suggestion applies to the patenting of the specific variety, as such. The more realistic commentators admit that most varieties will not qualify for patent protection because of the difficulty of showing that they entail an inventive step. Therefore there is no danger to the survival of the plant breeders' right as the preferred option for legal protection for innovations at the level of specific varieties.

The lack of examples of attempts to patent typical plant varieties tends to give this debate an academic rather than practical character, but the rejection by the Supreme Court of Canada of a patent application for a soya bean variety produced by methods of cross-breeding and selection (Pioneer Hi-Bred, 1988) provides a model of the type of patent claim that would be presented for a variety obtained in this way. The claim in this application was based essentially on a listing of phenotypical properties, and it might be difficult (as a general rule outside this particular case) to identify an inventive concept in any such property or in a combination. This concrete example should help in future to clarify the issues in discussions between patent and UPOV circles, which have hitherto often been at cross-purposes for want of agreement over terminology.

5. Patenting Plant Genes

In the area of gene isolation and cloning and the production of transgenic plants, the problems of public policy and possible legal conflict give way to the more familiar issues of patent law. These involve typical questions of novelty and inventive step in the light of prior knowledge and existence in nature and, although usually controversial in particular cases, are easier to resolve.

A good example of the type of patent granted in this field is the cloning of a DNA fragment originating in a micro-organism mutation which expresses a resistance to the herbicide glyphosate. Intended for integration into plants to confer glyphosate resistance, the gene is patented (in the USA) in terms of a DNA sequence containing the relevant structural gene. The latter is defined in broad functional language and, apart from a size restriction (less than 5 kilobases), is silent as to nucleotide sequence.

Increasing numbers of patents of this kind are now appearing, especially in the United States, and many applications are still in the procedural pipeline in the Japanese and European patent offices. The extent of the protection conferred by claims to DNA sequences is the subject of legal question in US law. In the pharmaceutical field, there is dispute as to the protection such claims give over the protein expression product. In the plant field, the question whether such DNA patents extend to plants transformed with the DNA should be easier to answer, although it may not be sufficient for commercial purposes to rely on them alone, without patents expressly covering the final plants. Biotechnology companies will want assurance that patents on the genetic material will not be devalued by particular restrictions, such as the application of an exhaustion of rights principle which would deny them an appropriate return on the uses of the DNA in the final marketed products.

6. Patents for Animals

As indicated previously, plants and animals are grouped together in patent laws and administrative rulings, and the question of patents on animals has therefore been touched upon above. There is no animal equivalent to plant variety protection under UPOV or national laws and consequently no problem of legal conflict or interface to contend with. Again, among the major patent systems, it is in Europe that the position is not totally clear-cut. In the onco-mouse application it has been confirmed by the Appeal Board that the exclusion of "animal varieties" from patentability under Article 53(b) must be narrowly interpreted and cannot apply to animals in general. However, the EPO were instructed also to consider the possible relevance of Article 53(a) (which bans patents for inventions the publication or exploitation of which would be contrary to public order or morality) and in doing so to balance the potential of the invention for human good against the possible risk of harm to the environment and the undoubted harm to the test animal. The EPO decided that in this case the benefit to cancer research outweighed the other factors, and the onco-mouse patent will therefore be granted. However, in another application involving a transgenic mouse in the screening of substances as hair-growth stimulants, the EPO are applying the test adversely to the applicant. Although the European Commission has supported the principle of patents for genetically engineered animals in its proposed Directive, the uncertainty in Europe for the many other applications awaiting outcome in the European Patent Office will persist at least until the moral criterion devised in the onco-mouse case is applied to them and the issue resolved case by case.

The above remarks apply to patents for genetically modified animals as such. As regards novel methods of cultivating and rearing animals, including fish, such methods seem to involve no more than straightforward issues of patentability, but patents are not yet numerous in this field.

7. Patents for Foods

Food biotechnology is a sector that has no special character in patent law outside the context of plants and animals. The patenting of food enzymes and novel foodstuffs, e.g. mycoprotein, is accommodated to the general rules of biotechnology patenting, although in some (mostly non-OECD) countries there is a general ban on patents for foods of any kind.

8. Opposition to Patents in Biotechnology

While the debate continues on the issues outlined above, a new tendency should be noted. Official patent circles and the industries that utilise biotechnology do not question the appropriateness of intellectual property for the new processes and products that emerge from the research and that show commercial promise. A highly vocal challenge to this assumption has come from animal rights and green movements and their supporters in the political arena and elsewhere.

Taking a stand on the alleged unethical practice of ''patenting life'', these groups often extend their opposition to any significant structural change in the agricultural industry that might arise from biotechnology and especially from the acquisition in the hands of multinational companies of monopoly rights on the advances that are being made. This argument is applied to both plant and animal biotechnology and in the latter case a moral objection is also raised against interference with the assumed right to integrity of the species. This opposition is targeted against the patenting of these inventions no less than against the research itself. The opposers have clearly appreciated the role of patent protection in stimulating the funding of this research, and their strategy is clear. This movement is highly active in the United States and in the European Parliamentary system and can be expected to maintain a high profile in public debate for some time to come.

This is of course, one point where the public acceptance and the patent issues could meet, although they are, as was explained at the beginning, of a very different order. Could opposition of the public or of minority groups, working through parliamentary and other procedures, spill over into the legislative and legal process and prevent the required modifications of patent law regarding biotechnology? Considering the changes which have already taken place and those which have been initiated, this possibility seems much less likely than in the safety and regulation context, although the opposition could cause further delay, particularly in Europe.

9. Further Reading

LESSER, W., ed. (1989), *Animal Patents: Legal, Economic and Social Issues,* Macmillan, London and Stockton Press, Cornell University Press, Ithaca, New York.

BENT, S.A., R.L. SCHWAAB, D.G. CONLIN and D.D. JEFFREY (1987), *Intellectual Property Rights in Biotechnology Worldwide,* Macmillan, London.

Commission of the European Communities (1988), *Proposal for a Directive on the Legal Protection of Biotechnological Inventions* – Comm., 496 final SYN 159, 17 October, Brussels.

CRESPI, R.S. (1990a), "Reflexions On the Onco-Mouse Decision (EPC)", in *Intellectual Property in Business,* Vol. 2, No. 6, pp. 25-26.

CRESPI, R.S. (1990b), "The Patenting of Genetic Resources", in *Impact of Science on Society,* Issue No. 158, Conserving and Managing our Genetic Resources, UNESCO, Paris.

European Patent Office (1987), *European Patent Convention,* 4th Edition, Munich.

OECD (1985), *Biotechnology and Patent Protection, an International Review,* Paris.

OTA (Office of Technology Assessment), (1989), "New Developments in Biotechnology 5: Patenting Life", Special Report OTA-BA-370, US Government Printing Office, Washington DC.

"Patenting Life Forms in Europe" (1989), *Proceedings of Conference in European Parliament,* Brussels, 7-8 February, ICDA Seeds Campaign, Barcelona.

STRAUS, J. (1987a), "The Relationship between Plant Variety Protection and Patent Protection for Biotechnological Inventions from an International Viewpoint", in *IIC (International Review of Industrial Property and Copyright Law),* 18, pp. 723-737.

STRAUS, J. (1987b), "The Principle of Dependence under Patents and Plant Breeders' Rights", in *Ind. Prop.,* pp. 433-443.

UPOV (1978), *International Convention for the Protection of New Varieties of Plants,* revised 19 March 1991, Geneva.

WIPO (World Intellectual Property Organisation) (1988), *Revised Suggested Solutions Concerning Industrial Property Protection of Biotechnological Inventions,* Document BioT/CE/1V/3 (WIPO), Geneva.

Patent Decisions mentioned in the text:
(USPQ = United States Patent Quarterly)

Chakrabarty (1980)	Diamond vs. Chakrabarty 206 USPQ 193.
Ciba Geigy (1984)	Official Journal of the European Patent Office 3/1984, 1/2.
Hibberd (1985)	Ex parte Hibberd 227 USPQ 443.
Calgene, Inc. (1985)	US Patent No. 4,535,060.
Quigg (1987)	D.J. Quigg, Commissioner of Patents, Official Gazette, United States Patent Office, 1077 OG 24, 21 April 1987.
Harvard (1988)	Onco-mouse US Patent No. 3,736,866.
Lubrizol (1988)	Lubrizol Genetics, Inc. Official Journal of the European Patent Office, March 1990, 59-62.
Pioneer Hybrid (1988)	14 Canadian Patent Reporter (3d) 491.
Harvard mouse (1990)	European Patent Office Decision, EPO Technical Board of Appeal Decision T19/90.

Chapter VIII

ECONOMIC IMPACTS

1. Introduction

a) *The framework for an economic impact assessment*

This chapter attempts to evaluate economic impacts of new biotechnologies in agriculture. The prospects and limits of such an attempt are conditioned by an assessment of the diffusion of biotechnology, which in turn depends upon various factors. The most important conditions for the diffusion of agricultural biotechnologies are: technical feasibility; economic profitability for the inventor and user; safety, related both to environment and human health; and acceptance by producers and consumers.

These conditions of diffusion are not independent but closely interrelated: technical feasibility of a specific biotechnology depends on the amount of investment in R&D, which in turn is influenced by expected profits. Low level of acceptance or lack of acceptable safety data might impede market introduction or sufficient market shares of an innovation.

The chapters in Part One were devoted to the potential of new biotechnologies from the perspective of science and technology. This survey revealed an overwhelming variety of technical options, many of which are already, or will soon be feasible. Also, many of these options could improve environmental safety or human health. Expected time scales for maturity of technologies might change due to unforeseen scientific developments and changes in government regulations, public acceptance and economic conditions.

The survey of company strategies with regard to new biotechnologies in upstream and downstream sectors of agriculture (Chapter V) can be summarised by the contradictory but complementary factors of interdependence, expressed in the globalization of the agrofood-complex, and heterogeneity, as seen in the development of many different markets. For this and other reasons, the economic profitability of many of the numerous scientific and technological options is neither obvious nor assured.

The review of public perception of new biotechnologies (Chapter VI) concludes that there are specific concerns, depending on the kind of biotechnology applied, the goals pursued, and the background of the persons asked. Lack of information is part of the acceptance problem. Lack of public confidence in regulations for new biotechnologies and food safety was identified as another part, fuelled by recent food scandals that were not related to new biotechnologies.

Economic impact assessments of new biotechnologies face a number of difficult methodological challenges:

i) Biotechnologies are not homogeneous but comprise a broad range of new options, varying in technical feasibility and profitability and with widely different economic impact potentials. Also, various biotechnologies are competing against each other in the same sectors.

ii) Agrofood biotechnologies pose problems of consumer acceptance, environmental and safety concerns and intellectual property protection. The economic internalisation of such external constraints has not received sufficient attention.

iii) Quality improvements, a key goal of agrofood biotechnologies, are not assessed in aggregate economic analysis and are not integrated into economic impact assessments.

The task of this chapter could be compared to an attempt to assess the impacts of chemistry or mechanics on agriculture before these techniques were introduced. Any visit to a technical museum shows that only a few routes of the numerous parallel technical developments prove to be successful in the long run. Sometimes, when social or environmental conditions change, some forgotten options meet with new interest and are selected for further development. To choose successful innovation strategies is the task of entrepreneurs in a market economy. It cannot be the goal of this economic impact assessment.

The following analysis will start with the changing role of agriculture and the food industry in OECD countries and the role of biotechnology for global food supply, in order to point to the major challenges for this sector in the years ahead. Two comprehensive studies on quantitative impacts of new biotechnologies on agricultural production in the USA and EC countries will be reviewed. This will be followed by a summary of more qualitative impact assessments on specific economic parameters, such as the structure of agriculture, prices of agricultural inputs and outputs, international trade and competitiveness, and employment. Then, case studies on bovine growth hormone are reviewed. Finally, a number of policy conclusions will be drawn.

b) The changing role of agriculture and the food industry in OECD countries

The role of agriculture and the food industry in society is changing, and new biotechnologies have appeared in a period of dramatic value changes. These shifts in values must play a major role in an impact assessment.

From a purely quantitative point of view, the importance of the agrofood industry has been declining with the economic growth of OECD countries. Due to natural saturation, demand for food is growing less than personal income, despite shifts to higher quality food products. This is often given as the main explanation – apart from high rates of labour substitution by capital – for the steady decrease of the percentage of total civilian employment in agriculture in OECD countries. In 1985, this percentage was 8.9 for the OECD area as a whole.

However, the decrease in agricultural employment must also be seen as part of a long-term trend whereby agricultural roles are increasingly taken over by manufacturing and service sectors. In other words, the decrease in agricultural employment indicates only to some degree a decrease in the macroeconomic significance of food. To a larger degree, it reflects a shift within the agrofood chain. As the numbers of farmers decrease over time, the importance of the food-processing and food-distribution sectors is growing, at least in relation to farming, but possibly also in absolute numbers, although this would require further statistical analysis.

Table 32.   **Employment in the agrofood sector in France, 1984**

	Absolute numbers of employed	Percentage of total employment
Agriculture, forestry and fishing	1 863 246	8.7
Manufacture of meat and dairy products	183 637	0.9
Manufacture of other food products	434 024	2.0
Wholesale food trade	289 901	1.4
Retail food trade	654 505	3.1
Hotels, cafes, restaurants	623 242	2.9
Agrofood sub-total	4 048 555	19.0
Total employment	21 349 459	100.0

Source: Les collections de l'INSEE, D 105 (May 1985), Survey of Employment for 1984.

Employment data for France may illustrate the structure of the agrofood chain in an industrialised country with a relatively large agricultural population (Table 32). Food-processing and distribution have more economic weight than farming. It should be noted that food distribution statistics are not comprehensive. If employment figures for food-transportation and various other food-related services could be added to the present figure of 7.4 per cent, employment in food distribution alone might match agricultural employment (8.7 per cent). Also, the processing and distribution of non-food products based on agricultural and forestry raw materials, such as wood and paper, are not included. These sectors would add a few percentages more to the employment figures. The relevance of these data in the context of this study is explained by the ability of biotechnology to intervene at every stage of the agrofood chain, from plant growing to food distribution, as has been demonstrated in earlier chapters.

Another way of statistically capturing the relative importance of food is to measure consumer expenditures on food, beverages and tobacco as a percentage of total household expenditure. Table 33 gives figures of EC countries in 1986. The percentage for all twelve EC countries is 21.9, with considerable variations between rich and poor countries. However, even in the richest countries, 17 to 20 per cent of total household expenditures are on food, beverages and tobacco.

In quantitative terms alone, the weight of the agrofood chain, although declining over time, is still substantial. However, agrofood issues are increasingly viewed with respect to other than purely quantitative criteria. Welfare societies show an increasing awareness of the environmental side-effects of technical and economic developments, and values shift from quantity to quality and to criteria that are not yet fully captured in economic terms.

Consumers are less concerned about food prices and sufficient food supply than about food quality (Chapter V). The term "food quality" is very heterogeneous and covers a wide range of almost contradictory trends: luxury foods and eating out in restaurants, highly processed convenience foods enriched with special flavours and vita-mins, and trends towards "natural food", such as that grown under the regulations of the International Federation of Organic Agricultural Movement. Many consumers have dif-

Table 33.   Consumer expenditures on food, semi-luxury food and tobacco in per cent of total household expenditures, EC, 1986

	Food	Non-alcoholic beverages	Alcoholic beverages	Tobacco	Total
Belgium	17.7	0.5	1.4	1.7	21.3
Denmark	16.4	0.6	3.6	3.0	23.5
Germany	12.7	0.5	2.2	1.6	17.0
Greece	33.0	1.4	2.6	2.9	39.9
Spain	24.5	0.4	11.1	1.3	27.2
France	16.8	0.5	2.1	1.1	20.5
Ireland	24.5	0.3	1.7	2.1	28.7
Italy	24.5	0.3	1.7	2.1	28.7
Luxembourg	14.9	0.5	1.5	6.4	23.3
Netherlands	14.8	0.5	2.0	1.8	19.1
Portugal	33.4	0.2	2.2	2.2	38.6
United Kingdom	13.6	0.6	1.9	2.8	16.9
EC-12	17.5	0.5	2.1	2.0	21.9

Source: Agra-Europe, German edition, 43/1990, 22.10.1990, Dokumentation, p. 1.

ferent eating habits, depending on working days, weekends or holidays, and cannot be strictly associated with just one of these trends.

However, except for qualitative descriptions and sensitivity analyses for specific biotechnologies, assessments of the economic impacts of quality factors have not been made, although much effort in biotechnology is concentrated on food quality.

Beyond the issue of food quality, the public is concerned about the quality of the production processes in agriculture and the effects on the natural environment: animal production conditions (keeping-systems, use of antibiotics and hormones, etc.) have been topics of public debate. Changes as well as legal restrictions have ensued.

Agriculture and forestry use major parts of the land in industrialised countries, e.g. 85 per cent in Germany. Farming practices therefore had an immense impact on landscape and natural environment in the past, and agriculture contributed to the beauty and variety of landscapes and thereby created a precondition for a greater variety of species. Due to the processes of growth, concentration, specialisation and intensification (increasing use of agrochemicals) in recent decades, the environment deteriorated: reduction of species, soil erosion, nitrate and phosphate leakage into ground and surface water, residue problems of pesticides in soil, water and food, erasure of biotopes and, possibly, contribution to the greenhouse effect are facets of the environmental problems caused by modern agriculture.

Although agricultural economists increasingly attempt to evaluate environmental impacts of agriculture and biotechnology impacts on agriculture, the economic assessment of the environmental benefits and costs of agricultural biotechnologies is an even more difficult task, which has hardly been addressed.

c) Subsidies and price policies at the crossroads: the GATT negotiations 1991

Agricultural policies that affect biotechnology impacts on agriculture may have entered a period of far-reaching change. So far, most OECD countries have protected their domestic farmers by price support, intervention and import restrictions for major agricultural products. As a consequence, agricultural supplies grew faster than demand. Increasing OECD surpluses have been sold on world markets at decreasing prices or given away as food aid. Accordingly, expenditures to support agricultural market and price policies expanded drastically.

Since approximately 1987, policy reforms have been initiated. These policy adjustments include reduced support prices for agricultural products, quota regimes, land-set-aside programmes, etc. However, no agreement has yet been reached in GATT negotiations (June 1991) on the extent and pace of reduction in agricultural protectionism.

Various efforts have been made to estimate the impacts of trade liberalisation. Trade liberalisation models differ, particularly with respect to theoretical and empirical content, base year, protection level, expected reduction of protection in the future and time horizon considered for adjustments. Hence, the estimated size of impacts can vary considerably (Dixit *et al.*, 1989; Hofreither, 1990).

Existing trade liberalisation models do not explicitly take biotechnologies and their prospective impacts into account. Moreover, the available studies of economic impacts of agrofood biotechnology neglect the fact that different agricultural policies could lead to different impacts. Only in studies of specific techniques, such as bST, have impacts been evaluated for different policy scenarios.

Two questions dominate a discussion of agricultural support policies in the context of an assessment of biotechnology impacts on the agrofood-chain: if trade is liberalised, will biotechnology alleviate or aggravate adjustment problems for the countries that enjoy substantial protection levels? Second, will a reduction of protection levels influence the rate, direction and composition of innovation and diffusion of new biotechnologies?

Looking at market surpluses and budgetary expenses in countries with high rates of protection, it can be argued that biotechnologies which increase productivity in agriculture by reducing inputs per unit of output are very desirable and would alleviate at least some adjustment problems. In contrast, yield-increasing impacts of biotechnologies will raise new concerns. These aspects are discussed in section 2.

A reduction of agricultural protection in OECD countries will certainly influence rate, direction and composition of innovation and diffusion of new biotechnologies. Although no explicit evaluation of this relation can be referred to, the theory of induced technical change suggests that the rate of innovation or diffusion will decrease in high-price countries or sectors if price levels are substantially reduced (e.g. milk or sugar in Japan, the EC and the United States), and increase in low-price countries if agricultural prices rise (e.g. "Cairns group" countries such as Australia and New Zealand). Still, the heterogeneity of biotechnology does not allow for many generalisations. Even with falling agricultural prices, some biotechnologies will stay competitive, while others will fail. What will certainly change is the direction and structure of innovation.

Additional factors have to be considered. Biotechnology will have direct and possibly stronger effects on the upstream and downstream sectors of agriculture than on agriculture itself, and the impacts of trade liberalisation on these sectors might be different from those on agriculture. Moreover, in the long term, the prospects might be very different from those prevailing in a short-term transition phase. The rate of structural change in sectors highly protected in the past will accelerate, and this in turn is likely to lead to efficiency increases and hence, faster adoption of biotechnological and other innovations. So far, interrelations between impacts of trade liberalisation and of biotechnology on the agrofood-chain have not been sufficiently and explicitly treated in the literature.

d) Biotechnology and long-term global food supply

Chapters I to IV, which reviewed scientific and technological options, have expressed the natural scientist's conviction that the contributions of biotechnology will be necessary to assure the long-term food supply for a growing world population. It is true that the role of biotechnology is now generally recognised to be important to achieve various, specific goals in the agrofood context, e.g. to reduce environmental damage due to agrochemical and other technologies, to increase the fuelwood supply of the Third World, or to improve food safety. However, the issue of whether and how the long-term global food supply might be linked to new biotechnology is not well articulated in the available long-term projections of supply and demand for basic foodstuffs.

On the whole, the projections by the World Bank, the FAO and the IFPRI (International Food Policy Research Institute, Washington) are relatively optimistic about continued productivity increases with current technologies. These projections assume a more general adoption of current "best practices", but it is not clear whether this includes biotechnologies, as specific reference to them is avoided.

These relatively optimistic projections imply a continuing strong increase in the production of basic foodstuffs, with OECD countries maintaining or increasing their exports, and food-deficient areas (e.g. Africa, southeast Asia, China) continuing to raise their own supplies and import their additional cereal requirements. They also imply that, at an aggregate level, sufficient calories will be available internationally to feed everyone relatively adequately, and that supply and productivity trends could maintain this position into the 21st century even with existing ("best practice") technology.

By contrast, two expert reports, reviewed below, specifically discuss the links between biotechnologies and agricultural productivity in the United States and in Europe. Both reach the conclusion that increases in agricultural productivity would soon level off without new biotechnologies, and have, in some cases, already done so. It must be emphasized, however, that these projections have been made for industrialised countries with advanced agricultural sectors and substantial surplus problems. Therefore, they cannot be extrapolated to other countries of the world or used to assert that the world's long-term food supply will depend on biotechnology, a question which is not directly addressed in either of the two reports.

Even if it were possible to provide enough food for a growing world population until some date in the next century by wider use of "best practice" technologies as well as by policy and institutional adjustments on a global scale, biotechnologies offer numerous additional options to reach the goal of an adequate food supply at lower cost including environmental costs. In other words, a scenario without biotechnology means more and faster use of "best practice" technology and more social and policy adjustment than would be necessary in a scenario with biotechnology. The first scenario, although not impossible, is not realistic, and certainly more costly than the second one.

The debate about whether the long-term response to global food needs should include biotechnology and if so, how much and what kind of biotechnology, applied to what purpose and to whose benefit, cannot be settled by arguments based on natural sciences or on food statistics alone. Even today, there is massive undernourishment in countries and regions that in principle could enjoy self-sufficiency in basic foodstuffs. The debate is not only about technology but also about economic, social and political forces and institutions. It is a debate about which goals in the agrofood sector could be reached primarily by economic, social and political change, which primarily by techno-logical change, and which are calling for complementary, socio-economic and technolog-ical solutions. It is also a debate about how, and for how long, technological change can be delayed or directed into other routes (e.g. integrated farming methods).

2. Impacts on Production and Market Balances

Economic impact assessments have focused most of their attention on impacts on agricultural productivity and production potentials. The main reason is that, economi-cally, such effects are easier to analyse than others. Effects on agricultural structures, trade and competitiveness can only be dealt with on the basis of estimated effects on production. This is why this section will first review production impacts.

Translated into economic language for an aggregate economic analysis, the numer-ous options offered by new biotechnologies can be classified into the following categories:

- *i)* Product innovations:
  - *a)* new or novel demand-specific products (high tryptophan corn, industrial soya, hard tomatoes with longer shelf-life);
  - *b)* alternative uses for existing products (bioethanol from grain);
  - *c)* substitutes for existing products (single cell protein).
- *ii)* Process innovations:
  - *a)* protection enhancers (diagnostic kits, vaccines);
  - *b)* performance enhancers (bST, *Rhizobia*);
  - *c)* improved traits in crop and livestock (resistance against diseases).

Categories *ia)* and *ib)* could create new market and income potentials for agricul-ture, whereas category *ic)* signifies new competition. However, Chapter V has already pointed out that OECD countries can expect new food markets only on the basis of quality improvements. New biotechnologies, such as bioreactor technologies, could lead to competition for existing food and feed products (category *ic)*, but at the present moment, this is likely only for very high-value, low-volume products, which in general concern cosmetics and pharmaceuticals more than food products.

Much greater opportunities for substitution appear to arise from process innovations (category *ii)*. Biotechnologies which improve processing could help reduce the costs of agricultural raw materials or substitute expensive materials by cheap materials. Perform-ance enhancers (*iib)* increase productivity and lower costs, protection enhancers (*iia)* reduce losses and thus stabilise and increase yields. Improved traits in crop and livestock (*iic)* can have very different impacts: herbicide resistant plants stabilise chemi-cal input into agriculture; pest resistant plants replace the use of agrochemicals. Some higher yielding varieties require increased use of fertilizers and often additional plant protection or adjustments in cultivation techniques, and enlarge the production capacity.

It has been argued that biotechnology's big potential should be used rather for cost reduction than for yield increases in order to reduce both agricultural surpluses and environmental damage. However, this strict distinction is hardly ever possible in practice: to be adopted, any new technology must be profitable. Any increase in profitability will induce farmers to extend production, unless quota systems restrict them. Nevertheless, breeding for resistance against pests and diseases could reduce the use of chemicals in agriculture and the related burden on the environment substantially, even if profitability is only slightly higher than for the chemical alternative. In this case, the resulting incentive to increase production would be marginal.

Although it would be impossible to analyse the numerous potential impacts in detail, various efforts have been made to assess the effects of new biotechnologies on agricultural productivity, production potentials and on market balances. However, among the reports available to a wider public, only two cover a wide variety of technologies or applications for larger markets of OECD countries. All other reports are more qualitative or focus on specific technologies or are limited to single countries. The two more comprehensive reports have been written for the United States Office for Technology Assessment and for the European Commission.

a) Quantitative results for the USA (OTA, 1986)

The Office of Technology Assessment of the US Congress (OTA) attempts to estimate in one chapter of its study on "Technology, Public Policy, and the Changing Structure of American Agriculture", yield and production effects of emerging technologies for various crops and animals. In a Delphi-survey, connected with workshops, estimates by experts were obtained for the timing of commercial introduction of the techniques, net yield increases, and speed of adoption. Tables 34 and 35 show the results. As the range of techniques covered include, in addition to biotechnology, changes in farming practices and information technology, the estimates have to be seen as synergistic effects or as upper limits for partial new biotechnology impacts.

Table 34. **Historical and projected rates of annual growth in US crop yields**

Per cent

| | 1962-82 | 1982-2000 | | |
		No-new-technology environment	Most likely environment	More-new-technology environment
Corn	2.6	0.5	1.2	1.6
Cotton	0.1	0.3	0.7	1.0
Rice	1.2	0.2	0.9	1.4
Soybean	1.2	0.8	1.2	1.2
Wheat	1.6	0.7	1.2	1.4

Source: Office of Technology Assessment (1988), p. 80, Table 3.4.

Table 35. US animal production efficiencies in 1982 and estimated annual growth rates

	1982 Level	1982-2000 Annual growth rates (per cent)		
		No-new-technology environment	Most likely environment	More-new-technology environment
Beef				
kg meat per kg feed	0.07	−0.3	0.2	0.2
Calves per cow	0.88	0.5	0.7	0.9
Dairy				
kg milk per kg feed	0.99	−0.2	0.2	0.6
Milk per cow per year (kg)	5 572	1.4	3.9	4.3
Poultry				
kg meat per kg feed	0.40	1.6	2.0	2.1
Eggs per layer per year	243	0.4	0.7	0.8
Swine				
kg meat per kg feed	0.157	0.4	0.6	0.8
Pigs per sow per year	14.40	0.5	1.1	1.2

Source: Office of Technology Assessment (1988), p. 80, Table 3.3.

Crop yields would increase further to the year 2000 even without new technologies due to lagging adjustment by late innovators. However, yields are expected to grow less than in the past in all three technical scenarios which have been chosen, except for cotton in general and rice in the scenario "more-new-technology environment". One reason is that marketable new biotechnologies in crop production are lagging behind those in animal production. Commercial introduction into the market is expected to start in the mid-1990s, full use of these techniques is projected for the first decade of the next century. The decline of yield increase in the "no-new-technology-environment" will be discussed below. Compared to these paths, yield increases can be substantially raised by new technologies as shown in the two columns to the right of Table 34.

Prospects differ for animal production: milk production per cow and year increased between 1960 and 1982 by 2.6 per cent on average. Poultry feed efficiency grew with an annual rate of 1.2 per cent before 1982. Beef and swine feed efficiencies, however, did not increase during the same period (OTA, 1986, text). For milk in the "no-new-technology environment" the annual yield increase is estimated to drop to 1.4 per cent (Table 35). Annual rates of up to 4.3 per cent are projected for the "more-new-technology environment", with a hint about the partial effect of bovine growth hormone (bST) alone in the range of 20 to 40 per cent, which equals an annual rate of 1 to 2 per cent for the analysed period. The bST technique will be specially treated below.

As general tendencies, the experts expect:

i) declining rates of growth in efficiency in the "no-new technology environment", compared to the past;

Table 36. **Projections of US crop production**

Crop	Unit	1984	2000 No-new-technology environment	2000 Most likely environment	2000 More-new-technology environment
Corn					
Production	Million tons	195	218	236	246
Growth rate	Per cent		0.7	1.2	1.5
Cotton					
Production	Million tons	2.81	2.90	3.13	3.26
Growth rate	Per cent		0.1	0.7	0.9
Rice					
Production	Million tons	13.7	15.4	16.3	16.9
Growth rate	Per cent		0.7	1.1	1.3
Soybean					
Production	Million tons	52	81.5	87	89.7
Growth rate	Per cent		3.1	3.4	3.6
Wheat					
Production	Million tons	71	90	95	95
Growth rate	Per cent		1.5	1.9	2.0

Source: Office of Technology Assessment (1988), p. 82, Table 3.5.

ii) substantial increases in growth of yield and production efficiencies by stepping from the "no-new-technology-environment" to the "most likely environment"; and

iii) additional positive but decreasing returns for a further move to the "more-new-technology environment".

As no reasons for the expert judgements have been given in the reports, it is not clear whether the experts expect declining rates of growth in efficiency in the scenario "no-new technology" for technical reasons only or also in response to surplus production and associated policy changes.

Expert estimates of crop yields and crop areas planted, as well as of improvements in animal production, were used in an econometric model of the US agricultural sector for a 20 year simulation on aggregate production impacts. The results are shown in Tables 36 and 37. Crop production increases for corn, cotton and rice at similar rates as the respective yields, because the areas planted remain constant. For soybeans and wheat, however, production increases faster than yields because of increases in acreage.

Higher production of crops leads to lower prices and lower feeding costs. Reduced feeding costs together with shifts in demand and technology changes in animal production lead to rising production of poultry in all technical scenarios and for milk in the scenarios with new technologies. Whether the high rate of annual milk production increases is realistic, or merely an effect of model parameters, cannot be judged from the

Table 37. **Projections of US animal production**

Livestock	Unit	1984	2000		
			No-new-technology environment	Most likely environment	More-new-technology environment
Prime beef					
Production	Million tons	7.57	5.66	6.39	7.11
Growth rate	Per cent		−1.5	−0.8	−0.2
Poultry					
Production	Million tons	6.12	7.61	7.57	7.57
Growth rate	Per cent		1.4	1.3	1.3
Pork					
Production	Million tons	6.25	4.85	5.30	5.89
Growth rate	Per cent		−1.6	−1.0	−0.4
Milk					
Production	Million tons	61.3	57.1	87.0	91.4
Growth rate	Per cent		−0.4	2.2	2.5

Source: Office of Technology Assessment (1988), p. 84, Table 3.6.

report. Prime-beef and pork production are declining, but the extent of this decline is reduced with increasing adoption of new technologies, which in turn allows for price cuts and corresponding increases in demand.

North American grain exports are projected to double by the year 2000. This result, however, has to be seen as a potential effect; its realisation depends on the external forces of world markets and future policies as well as on the internal flexibility of the price system.

b) *The impact of biotechnology on agriculture in the European Community to the year 2005 (Neville-Rolfe et al.)*

No study comparable in scope, endeavour or methodology to the OTA study exists for the European Community. However, Neville-Rolfe *et al.* (1988) carried out a Delphi-survey for the EC, along lines quite similar to those chosen in the OTA analysis. As the time perspective is 2005, a higher rate of diffusion of new biotechnologies should be expected, compared to the OTA estimates for the year 2000. For major crops and animals, the impacts of various new biotechnologies were asked for separately. The impacts of chemical growth regulators on cereals (Table 38) explains how these calcula-tions are made.

Adoption rates and expected increases in crop yields or livestock production are derived from a Delphi-process. Expected uptake of chemical growth regulators was estimated to concern 33 per cent of cereal producers, and yields were expected to increase by 10 per cent. Assuming that the largest producers are also the first innovators, the proportion of land or herds affected are calculated with 1985 data for farm size and

Table 38. **Estimated impacts of various new biotechnologies on different crop productions to the year 2005 in the European Community**

Per cent

	Cereals	Sugar	Oilseeds
Chemical growth regulators			
Expected uptake by 2005	33.0		
Area equivalent	80.0		
Expected yield improvement	10.0		
Increase in production possible from use of chemical growth regulators	8.0		
Average annual rates 1985-2005	0.4		
Stimulatory micro-organisms			
Expected uptake by 2005	15.0	37.0	34.0
Area equivalent	61.0	74.0	74.0
Expected yield improvement	8.0	9.0	10.0
Increase in production possible from use of stimulatory micro-organisms	4.9	6.7	7.4
Average annual rate 1985-2005	0.2	0.3	0.4
DNA/MAB probes			
Expected uptake by 2005	67.0	72.0	72.0
Area equivalent	94.0	94.0	93.0
Area treated	47.0	47.0	47.0
Expected yield improvement	12.0	14.0	17.0
Increase in production possible from use of DNA/MAB probes	5.6	6.6	7.9
Average annual rates 1985-2005	0.3	0.3	0.4
Transformation of crops with genes for viral components			
Expected uptake by 2005	48.0	36.0	30.0
Area equivalent	90.0	73.0	70.0
Expected yield improvement	12.0	5.0	5.0
Increase in production possible from use of transformed crops with genes for viral components	10.8	3.7	3.5
Average annual rates 1985-2005	0.5	0.2	0.2
Broad spectrum herbicide resistance			
Expected uptake by 2005	48.0	62.0	50.0
Area equivalent	90.0	84.0	86.0
Area treated	45.0	42.0	43.0
Expected yield improvement	8.0	11.0	11.0
Increase in production possible from broad spectrum herbicide resistance	3.6	4.6	4.7
Average annual rates 1985-2005	0.2	0.2	0.2
Incorporation of pest-combatting genetic characteristics			
Expected uptake by 2005	36.0	38.0	32.0
Area equivalent	84.0	76.0	73.0
Expected yield improvement	7.0	8.0	7.0
Increase in production possible from incorporation of pest-combating genetic characteristics	5.9	6.1	5.1
Average annual rates 1985-2005	0.3	0.3	0.2

Source: Neville-Rolfe *et al.* (1989), p.63 ff.

Table 39. Estimated impacts of various new biotechnologies on different livestock productions to the year 2005 in the European Community

Per cent

	Livestock Production			
	Dairy Cattle	Beef Cattle	Pigs	Sheep
DNA/MAB probes				
Expected producer uptake by 2005	65.0	65.0	66.0	55.0
Proportion of herd affected	95.0	95.0	99.0	87.0
Performance improvement expected	7.0	8.0	7.0	5.0
Potential impact of DNA/MAB probes on output to 2005	6.7	7.6	6.9	4.4
Average annual rates 1986-2005	0.3	0.4	0.4	0.2
Recombinant viral vaccines				
Expected producer uptake by 2005	67.0	65.0	68.0	50.0
Proportion of herd affected	96.0	95.0	99.0	81.0
Performance improvement expected	12.0	12.0	12.0	7.0
Potential impact of recombinant viral vaccines on output to 2005	11.5	11.4	11.9	5.7
Average annual rates 1986-2005	0.6	0.6	0.6	0.3
Embryo transfer, transgenic animals, endocrine interference and immunochemical control				
Expected producer uptake by 2005	>50.0		>50.0	>30.0
Proportion of herd affected	90.0		98.0	75.0
Performance improvement expected	10.0		10.0	12.0
Potential impact of above technologies on output to 2005	9.0		9.9	9.0
Average annual rates 1986-2005	0.5		0.5	0.5

Source: Neville-Rolfe *et al.* (1989), p. 66 ff.

1986 data for herd size structure. The largest cereal producers (33 per cent) controlled 80 per cent of the cereal production area. The production impact is derived as the product of 80 per cent of area equivalent times 10 per cent of expected yield improvement. As farm and herd size will substantially increase by the year 2005, the proportion of production affected will accordingly be underestimated. Results are shown in Tables 38 and 39.

As the authors point out, the various estimates for each group of new biotechnologies cannot simply be added up, because cross-impacts may exist, and comprehensive effects of all biotechnologies together have not been investigated. It must also be mentioned that bST was excluded from the analysis of dairy cattle. The estimates must be interpreted as additional impacts of new biotechnologies coming on top of the long-term productivity increases due to traditional factors such as conventional breeding, delayed adoption of existing techniques etc.

During the period from 1973 to 1985, annual yields for major crops increased by rates of approximately 1.5 per cent for potatoes, and approximately 3.6 per cent for soft wheat in the EC. For special crops, significantly higher levels have been obtained.

Without the use of biotechnology, the authors expect reductions of these rates by up to 30 per cent, depending on the crop, for the period of projection. Biotechnologies, however, could even lead to higher rates of growth than in the past. In plant production, high adoption rates are expected for DNA/MAB probes, medium and low rates for other technologies. Some individual groups of comparable biotechnologies are estimated to be able to increase average annual production by up to 0.5 per cent merely by increasing yields.

For livestock production, the effects of individual groups of biotechnologies are estimated to be slightly higher than for plants. The strongest impact on production is expected from recombinant viral vaccines, followed closely by other technologies. The effects that differences in productivity increases between various crops and animals will have on production patterns, total production and prices, have not been studied. Hence, the results represent expectations regarding the technical potential of new biotechnologies.

c) Discussion and conclusions

A few additional studies of a similar kind, but narrower with regard to range of biotechnologies, production lines or geographic regions examined, have been carried out. All available studies have shortcomings which have to be kept in mind when results are interpreted. The following limitations are important:

i) Some of the techniques will face delays in realisation. Unexpected technical breakthroughs or obstacles might change time patterns of technical feasibility.

ii) The reviewed quantifications are restricted to the estimated quantitative effect on yield and production, whereas many new biotechnologies aim at input reduction or quality improvement in production processes or in products. Moreover, while some qualitative changes will be incorporated in prices, no attempts have been made to estimate such quality-induced price effects on aggregate levels. Anyway, such estimates could be made only for specific techniques or products and when sufficient data are available.

iii) Costs of new biotechnologies have not been taken into account.

iv) Costs for compliance with safety regulations might add to the costs.

v) Lack of acceptance of certain new biotechnologies could ban them from the market, if political decisions against approval are taken, or could severely limit their adoption, if the decision is left to producers and consumers.

Comparing the results of the EC and the OTA reports requires considerable caution, because the historical and the future time-horizons of both studies are different and because EC farm structures deviate substantially from those in the USA.

Nevertheless, a number of common points and conclusions have emerged from these studies:

i) Without biotechnologies, yield increases will level off in the future compared to historical rates. Growth rates have already begun to decline in the last years.

ii) Biotechnologies incorporate a substantial potential for yield increases. However, the net impacts on agriculture will not be as big as originally expected, because biotechnology will partly make up for the declining trends in yield increases resulting from traditional technical factors.

iii) In the USA, the experts involved in the Delphi process expect crop yield increases in the future to be smaller than those in the past, even in the scenario with more new technology. Although this question was not raised explicitly in the European study, partial effects estimated for specific biotechnologies reach a size which indicates that the sum of biotechnology effects might be larger than the expected reductions in traditional yields increases, in contrast to the forecast for the USA. However, this difference may partly be due to the longer time horizon of the EC study, compared to the OTA report.

iv) Uncertainty prevails with regard to synergisms between different biotechnologies, e.g. between yield enhancers and breeding technologies.

v) If all quantitative trends develop as expected, the potential for further surplus production in OECD countries will increase unless countervailing policy options are implemented, e.g. price cuts in the framework of GATT negotiations, quota systems, land set-aside programmes, extensification programmes, early retirement schemes for farmers, etc. In other words, whether or not biotechnology leads to quantitative increases is more a question of policy than of technology.

An assessment of other factors such as price, income or structural effects will necessarily be influenced by the fact that estimates of quantitative production effects of biotechnology have still very limited reliability. Thus, an analysis of such further economic effects calls for an even bigger note of caution because of the additional uncertainties linked to political and economic forces. The following sections will therefore present largely qualitative or comparative assessments, as there are hardly any quantitative estimates other than those on production impacts.

3. Impacts on Agricultural Input and Output Prices

Two perspectives are possible: one regards biotechnologies as one form of technical change among others, and one looks at specific impacts of certain biotechnologies.

a) *General effect of technical progress on agricultural input and output prices*

From a farmer's perspective, biotechnology, like any other source of technical change in a market system, works as a "treadmill" (Cochrane, 1958): in order to be adopted, a new technique must offer an economic advantage to the farmer beyond the risk associated with its introduction. Risk-taking farmers are the early adopters. If the introduction of the new technique is successful, adopters receive an extra profit or pioneer rent. More cautious farmers copy the practices of early innovators. As time proceeds, the innovation spreads in the sector. Improved profits lead to higher supply, which in turn induces falling product prices. Late innovators are forced to adjust or to bear the pressure of reduced profits. The diffusion of the innovation follows the well-known S-shaped curve upwards over time, reaching a final saturation level. Pioneer rents decline in an S-shaped curve downwards. Fading extra profits induce innovative farmers to start searching for new technical changes and rents: a new cycle starts and the treadmill character of technical change emerges again. This mechanism can be seen as the innovative engine of market economies.

Pioneer rents are distributed in the market system among participants in the agrofood chain (e.g. input suppliers, farmers, traders and processors, final consumers) according to their relative market powers. Under conditions of competition the benefits finally accrue to the consumers. The pressure of permanent adjustments on producers and the advantage to consumers of having plentiful products at low prices is the Janus-face of technical change in flexible market systems.

Three trends distinguish agriculture in a free market system when technology-induced price effects make themselves felt:

i) concentration in upstream and downstream sectors;

ii) processing, trade and distribution of food are increasingly transferred from agriculture to downstream sectors;

iii) structures in agriculture become competitive, with high rates of innovation and decreasing price responsiveness of demand for agricultural products.

Accordingly, a relatively high proportion of innovation rents in agriculture would, under these circumstances, be transmitted by the price mechanism to other groups in the agrofood-business and to consumers.

However, in the markets for agricultural products regulated by government intervention in the form of price support policies and related measures, which are typical for major products in many OECD countries, the price cuts in response to technical change accrued in the past only partly to consumers. Real prices for agricultural products declined, but at a lower rate than would have been possible under market conditions. Two major unintended side-effects occurred. First, a part of the increased profitability of farming due to agricultural price support policy was captured by owners of fixed sources of production (land, quotas). Furthermore, production increased faster than demand and thus, surpluses piled up, which were dumped on world markets.

In the future, two different scenarios for agricultural price policies in industrialised countries have to be considered as a possible framework for the diffusion of new biotechnologies: continued price support with quantity controls via quota systems and land set-aside schemes, or a stepwise reduction of agricultural protectionism with financial transfers for adjustment and, perhaps, income support.

b) Specific effects of new biotechnologies on agricultural input and output prices

As stated earlier, no generally valid statement for all new biotechnologies can be made. However, a number of reasons argue for the potential of new biotechnologies to permit price cuts for agricultural inputs:

i) Substitution of old methods of input production by new biotechnologies, (e.g. tissue culture in plant breeding to produce homozygous plants) could substantially reduce breeding time and costs.

ii) New substitutes for farm inputs, such as insect- or pest-resistant plants to replace insecticides or pesticides, could increase competition and therefore reduce input prices.

iii) New biotechnology firms entering the input supply market could also increase competition.

On the other hand, concerns have arisen that the suppliers of agricultural inputs will try to capture as much of the pioneer-rent as possible, by improved property rights for plants and animals (patents) and by mergers and alliances between new biotechnology firms and traditional input suppliers (e.g. agrochemical and seed industries interested in herbicide resistant plants). As a result, the farmer would have to buy a package, including the herbicide-resistant plant and the associated herbicide.

The prospects of such structural changes in the seed and agrofood industries have been analysed by various authors and are also referred to in Chapter V. As the new products are not yet on the market, and as competitive behaviour in oligopolistic markets cannot be deduced theoretically, only some general, limiting conditions for the impact of biotechnologies on agricultural inputs price can be mentioned. A number of factors militate against high prices for farm inputs:

   i) a clear economic advantage for the farmer is required to adopt innovations;
   ii) lower input prices increase demand, because more farmers can be induced to buy the new factor of production, whereas higher input prices reduce demand;
   iii) high innovation rents would induce new competitors to enter the market and search for substitutes.

As to possible aggregate effects on output prices, three structural impacts of new biotechnologies on agricultural production must be analysed:

   i) new markets for agricultural products could be created, particularly in the field of non-food and renewable resources, but also for food and feed products (tomatoes with longer shelf-life, tryptophan corn);
   ii) alternative uses for existing products could be developed, e.g. bio-ethanol from grain;
   iii) new competition for existing products by substitutes such as single cell protein, etc., will emerge.

Hopes for agriculture are especially based on expanding substitution of non-renewable by renewable agricultural resources (see Chapter IV). If biotechnologies could bring about more efficient uses of non-food products from agriculture compared to food products, the fact that non-food products do not face the same inelasticities of demand as food products could contribute to maintaining relatively higher prices for agricultural products. Although this option cannot solve current surplus problems in OECD countries, it will be relevant in the long run. Currently, two major price distortions limit this option: agricultural protection in OECD countries hampers the shift of agricultural resources from food to non-food production and prices of non-renewable resources do not reflect long-run scarcities. From an economic point of view both distortions should be corrected.

For food products, the treadmill mechanisms will work with modifications by policy intervention. The use of biotechnologies for quality improvements of food products will have a positive impact on the food industry, but farmers working in contract agriculture might benefit as well, if their bargaining position is strong enough. Increased substitutability of agricultural raw materials would, however, have a depressing effect on agricultural product prices, although replacement of farm products by non-agricultural substitutes is not seen as a problem for OECD countries at the present time.

One of the consequences of increased concentration in upstream and downstream sectors of agriculture is likely to be the need to better monitor concentration processes and to strengthen the economic bargaining position of agriculture. Finally, one must mention that price policy exerts an influence on the level and direction of technical change. This important feedback from price policy to biotechnologies, however, cannot be discussed in this framework.

4. Impacts on Farm Structure

a) General framework

In public and scientific discussions about agriculture, the term "structure" can be related to very different characteristics or components: size and number of farms, specialised versus mixed farms, owner versus rental farms, family versus industrial farms, part-time versus full-time farms, etc. In the past the process of change in the structure of agriculture has been characterised by a decrease in the number of farms, an increase in the size of surviving farms, growing specialisation, concentration, commercialisation and division between ownership and management, as well as increasing linkages with other sectors of the economy. The complexity of this process of change and its speed have been reasons for public concern.

How will new biotechnologies influence this process of change? Scientific treatment of this topic has been even more sparse than assessments of production impacts. Structural analyses so far have focused more on changes in upstream sectors, such as ties between universities and industries or co-operation and mergers between new biotechnology, seed, and agrochemical firms (Kloppenburg, 1988; Joly, 1989). This is no surprise, as in these sectors, changes due to biotechnology have already occurred, partly in anticipation of expected future impacts, whereas the consequences for the downstream agricultural sector will occur only in the future.

The most comprehensive analysis of impacts on farm structures and regions was presented in the already reviewed OTA report (1986). Based on a summary of this analysis, and on qualitative assessments by other sources, some conclusions can be drawn.

b) Impacts on the structure of American agriculture (OTA, 1986)

The core of OTA's analysis can be described as an evaluation of cross-effects among developments in technology, public policy and agricultural structures. With regard to structural change, the evaluation proceeds by the following steps:

i) description of changes between 1969 and 1982 and of their determining forces;
ii) projection of changes for the year 2000;
iii) impacts of emerging technologies on structural change;
iv) impacts of structural change on the development and adoption of technical change.

Past changes in the structure of agriculture have been shaped by agro-economic factors (particularly concentration and vertical integration), technological factors (economies of scale, specialisation, capital requirements), institutional factors (research and extension), public policy (commodity programs, tax policy) and finally, macroeconomic developments (particularly employment opportunities outside agriculture).

Based on trends between 1969 and 1982, the total number of farms is projected to decline from 2.2 million in 1982 to 1.2 million in the year 2000. The number of large and very large farms with sales of more than $200 000 will increase, and 4 per cent of farms with sales of above $500 000 will control 75 per cent of all farm sales. At the other end of the spectrum, small hobby, recreation and part-time farms are rated as fairly viable and stable. Most pressure to adjust in either direction is expected for medium-sized farms with sales figures of between $100 000 and $200 000. As these medium-sized farms are seen as the backbone of US agriculture and as the focus of the public farm crisis debate, this conclusion has raised controversies (Thompson, 1990).

The major findings indicate that:

i) The rate of structural change has slowed down in the 1980s compared to the 1970s, mainly due to recession and increasing levels of unemployment. Parallels exist in European countries. Projections by the US Department of Agriculture based on these trends have led to forecasts of lower levels of change for the year 2000.

ii) The gradual disappearance of today's small and medium-sized, full-time farms can be interpreted differently when more size classes are distinguished: the average size of viable small and medium-sized farms in the year 2000 will be larger than today.

iii) Sales figures are a crude indicator for farm viability. Management quality and specialised strategies to survive in natural and economic niches can, to a certain extent, substitute for small farm size.

The structural impacts of biotechnologies have been analysed by experts in a Delphi process. This includes co-ordination and control, market access, and effects on labour and capital input.

It is shown that vertical co-ordination between agriculture and upstream and downstream sectors and control of agriculture will be intensified in response to some new biotechnologies. The influence of plant biotechnologies in this regard, however, is estimated to be only slight. In contrast, no impact on co-ordination and control is expected by mechanical and information technologies. Closer vertical co-ordination and integration will shift control over production from the farmer to the integrating forces in upstream or downstream sectors; this is similar to experiences of the US poultry industry in the past. Control will extend to production processes, not only to products. Another important effect of increasing vertical integration could be to further stimulate the concentration of production capacities in fewer, but larger production units.

Direct market access for animal products is expected to decrease slightly as a result of the new biotechnologies and of the ensuing increase in vertical integration. Farmers might find their market choices reduced. For plant products this effect is considered to be negligible. Also, new biotechnologies are not expected to significantly affect barriers to entry in agricultural production.

Impacts of new biotechnologies on the input of labour and capital, including land, are discussed on a general level. It is estimated that the amount of labour used in animal and plant production will be reduced by less than 5 per cent. Future capital requirements in animal production are also estimated to decrease slightly in response to the use of growth hormones and to new reproduction and genetic engineering techniques. In plant production no influence of new biotechnologies on capital input is anticipated. In sum, the impacts of new biotechnologies on capital and labour inputs into agricultural production are rated as small, and the capital-labour ratio is not expected to be changed significantly. It must be kept in mind, however, that the structure of American agriculture is different from that of most other OECD countries, and that its development level is higher. A similar study for European countries, or a study with a time-horizon beyond the year 2000, might have led to different conclusions regarding the potential factor-saving effects of agricultural biotechnologies.

Biotechnologies will exert a direct influence on adopters and an indirect one on non-adopters through increased competition. Adoption rates have been estimated, broken down into size (sales categories) and technology groups. The experts agreed on two major tendencies: adoption rates will be higher in the year 2000 for new biotechnologies than for mechanical and information technologies, and adoption rates will increase with the size of sales. The reason for the first finding is that most new biotechnologies are expected to arrive to the farm-gate as embodied technical change, that is as improved seeds or animals, or that they will be delivered as a specialist service. As this forecast applies more to plant than to animal agriculture, adoption rates for plant biotechnologies will be higher than those for animal biotechnologies. In contrast, information technologies and mechanical innovations require higher amounts of capital, which explains why the experts expect lower rates of adoption for those technologies. Thus, technology assessment appears to confirm the general, though only empirical impression of scientists who see agricultural biotechnologies gradually becoming more important in comparison to other agricultural sciences and technologies.

For similar reasons other studies have classified biotechnologies as technologically scale neutral, at least in principle. Nevertheless, adoption rates are expected to increase with farm size (measured by income potential, not acreage). The reason is that management quality is seen as a major force for adoption. As management quality tends to increase with farm size, higher biotechnology adoption rates could be found in larger farms as indirect effect. While other qualitative studies have pursued the same line of argument, critics have asked for more empirical proof and have particularly questioned the assumption that the largest farms are truly the most innovative ones.

The reverse impact, that of structural change and increased concentration on the rate of development of agricultural biotechnologies, is considered to be small. This conclusion results from the assumed technological scale-neutrality of biotechnologies.

The structural impact issue would be different if it were disaggregated into specific techniques, regions and products. The OTA study gives examples of some specific and more substantial impacts.

Other studies on specific structural impacts have been written for some groups of techniques or for specific countries. However, no generally valid conclusions can be drawn, although one comment is possible: in a closed market, a higher rate of technical change will induce increased rates of structural change, whereas in open economies, it will improve the competitive position of the technological leaders. OECD countries with viable biotechnological and agricultural sectors will tend to emphasise the second strategy, which constitutes the international dimension of the treadmill character of technical change.

5. Impacts on International Competitiveness and Trade

One might distinguish three different trade flows that will be influenced by the use of new biotechnologies: trade between OECD countries, between OECD and former Eastern bloc countries, and between OECD and developing countries.

Table 40 shows shares of the world's farm product exports contributed by these three trading groups. Developing countries export mainly raw materials, and many of them depend on one or two major products. Industrialised countries export animal products, food and feed grains, among others.

These trade flows have been heavily influenced by political intervention, as exemplified by protection of domestic agriculture in industrialised countries, taxes on agricultural exports as a source of government funds in developing countries, bilateral preferential trade agreements and economic planning in former Eastern bloc countries. Transition to market systems in the latter group and GATT negotiations might substantially change trade flows and international competition and hence the framework for research, development and diffusion of biotechnologies. As has been pointed out, the potential cross-impacts between trade liberalisation and biotechnologies have not been sufficiently evaluated.

The role of government support in the evolution of biotechnology and the impact of this support on competition between countries have been discussed in an earlier OECD study (*Biotechnology and the Changing Role of Government,* 1988). Improving the international competitiveness of US agriculture by fostering the US lead in agricultural technology was an explicit aim of the OTA report discussed above. In contrast, trade and competitiveness between industrialised and former Eastern bloc countries have hardly been treated in professional discussions. The major analytical emphasis has been on the ties between industrialised and developing countries, for various reasons: industrialised

Table 40. **Trade in farm products**

Product Group	Percentage export share (1986)			
	Industrialised countries	Developing areas	Former eastern bloc	Total world export
Dairy products	95.2	1.7	2.9	100.0
Beverages excluding tropical	87.9	5.2	6.9	100.0
Pulp	87.7	7.1	5.2	100.0
Hides and skins	86.2	6.9	5.7	100.0
Cereals and cereal preparations	82.5	12.1	5.7	100.0
Meat and meat preparations	78.2	11.4	10.3	100.0
Tobacco and tobacco products	70.2	21.3	8.5	100.0
Crude materials	68.5	22.5	9.0	100.0
Wood	65.3	21.9	14.1	100.0
Fruits and vegetables	61.5	30.7	7.2	100.0
Oilseeds, fats, oils, oilcakes, etc	61.4	32.4	6.4	100.0
Fish and fish preparations	57.7	35.8	6.5	100.0
Natural fibres	48.7	30.9	20.4	100.0
Sugar	23.3	71.3	4.0	100.0
Tropical beverages	9.4	88.7	1.9	100.0
Natural rubber	2.6	96.3	1.5	100.0

Source: Barker (1990), p. 306, based on GATT statistics.

countries have been the major driving force in biotechnology R&D, the substitution of sugar by high fructose corn syrup (HFCS) has demonstrated the possible magnitude of biotechnology-induced changes for developing countries, and finally poverty and hunger in the southern hemisphere are receiving more and more attention.

a) Technological substitution potential in OECD countries for developing countries' crops

Technically, three effects of biotechnologies on agricultural production and trade can be expected for plant production, the ultimate source of all food and feed and the major source for the poor (Junne, 1990):

   i) separation of plants from their original geographical environment, due to new plant qualities, such as higher tolerance to environmental stress (e.g. maize growing areas extending into northern zones);

   ii) separation of plants from their original characteristics, due to new processing of agricultural raw materials (e.g. HFCS produced from maize);

   iii) separation of the production of plant components from agricultural land, due to production by industry and substitution (e.g. fragrances and flavours).

In economic terms the result of these trends can be expressed as expansion of the potential production area suitable for plants and increasing competition between plants and/or between regions; increasing substitutability of and competition between agricultural products and new markets for non-food products from agriculture; industrial substitution of agricultural products.

Increased production areas, which on a global scale would have to be balanced against losses due to environmental degradation, and supply of non-food products will enlarge agricultural production opportunities, whereas industrial substitution of agricultural production will reduce markets for agriculture.

The main common impact of all three trends will be increased substitution and competition among agricultural products and regions. In principle, the technical potential for substitution could benefit developing countries as well. In reality, trade policies and unequal access to biotechnologies will determine substitution and the resulting impacts on trade.

b) Main findings on trade impacts in an OECD report of 1989

Present and currently predictable agricultural trade impacts for OECD and developing countries have been analysed in another OECD report (*Biotechnology: Economic and Wider Impacts*, 1989).

Three different cases, based on new technical options, are presented for four fields of application. These three cases have been widely mentioned in the economic literature as indications of a potential threat to developing countries. Hence, it is important to understand their limitations and to follow how developments between 1989 and 1991 have modified earlier conclusions and generalisations regarding substitution effects.

The cases are:

   i) Enzyme-based extraction of a liquid sweetener from maize as a substitute for beet and cane sugar, called high fructose corn syrup (HFCS) in the United States and isoglucose in Europe. The worldwide consumption of HFCS increased from

500 000 tonnes of raw sugar equivalent in 1975 to nearly 6 million tonnes in 1985, with annual growth rates of almost 20 per cent between 1980 and 1985. However, this was due to trade policies as much as to the new technical substitution potential, as will be shown in section c.

ii) Single-cell proteins produced by mass culture of yeasts or bacteria from hydrocarbons or other industrial substrates. The aim of single-cell proteins was to replace agriculturally produced protein animal feeds, but the much lower costs of soya and regulatory hurdles resulting from health effects of single-cell protein production have been major hindrances to market access in OECD countries. Although there may still be opportunities for oil-exporting countries on the basis of "free" gas, the single-cell protein case is an example of a substitution failure, at least in the short run.

iii) *In vitro* plant propagation and cell tissue culture, either to improve plants such as palm trees, or to substitute agricultural commodities, such as cocoa butter, pharmaceuticals, dyes, flavourings or fragrances by industrially produced inputs. *In vitro* cloning has improved palm trees with respect to productivity and easier harvesting conditions. The acreage planted with these improved trees is still so marginal compared to total world acreage that no influence on competitiveness and trade can be felt. World demand for oils has grown rapidly, with output of palm and cabbage palm oil doubling between 1960 and 1980 and continuing to increase. However, it is not an industrialised but a developing country, Malaysia, that has emerged as the dominant producer and exporter and that might gain most from the development of biotechnologically improved trees as well. Simultaneously, the oil and fat processing industries will find new substitutes for raw materials (by other than *in vitro* technologies), and will thus strengthen their position with regard to agriculture.

Plant cell tissue culture can also be used to substitute for agricultural products. Cocoa butter was mentioned because of all commodities for which biotechnological substitution R&D is carried out, it is the one with the largest market (almost US$ 900 million world market in 1985). As R&D costs are high, research concentrates on high-value, low-volume products, considered to be the only sector where substitutes will be economically profitable. Compared to global food markets, such substitutions are currently expected to be of very minor weight, although for individual countries and producers affected, they could imply losses of important production opportunities. In the very long term, however, plant tissue culture is likely to have far greater substitution potential than synthetic replacement of agricultural raw materials, because the first is generally, the second only selectively, applicable.

The picture of expected trade impacts between industrialised and developing countries due to biotechnology has changed in two ways since 1989: the discussion of trade impacts is now more detailed, and the economic prospects of biotechnology are considered to be more modest in 1991 than in 1989.

Impact assessments are becoming increasingly differentiated. Net trade position, agricultural and trade policies, potential for R&D and diffusion are major determinants which must be distinguished (Junne, 1990). Even for a single market, such as sugar, the potential influences of biotechnologies are heterogeneous and difficult to monitor (Knerr, 1990). This is a new challenge for agricultural market analyses, with a substantial degree of uncertainty.

Also, biotechnology development and diffusion is now seen as an evolutionary rather than a revolutionary process, although this might be different for some specific techniques. The likely time horizons for commercial introduction have been extended. Major impacts on plant production are not expected before the next century even in industrialised countries. In developing countries, there will be additional time lags before significant production impacts of new agrobiotechnologies will appear.

c) Trade shifts as a result of trade policies

The example of HFCS, which is so popular in substitution discussions, demonstrates the power of trade policies probably more than that of biotechnology. In 1985, the United States consumed 73 per cent of the total world production of HFCS, Japan 11 per cent, Europe 5 per cent and Canada 3 per cent. These different levels of diffusion of HFCS are closely related to agricultural and trade policies. The US government protects domestic sugar producers by import restrictions and guaranteed prices that most often substantially exceed the strongly fluctuating world prices. HFCS has become profitable as an unintended side-effect. However, with estimated production costs of $0.15 to 0.22 per pound, as against estimated long-term world sugar prices of $0.12 to 0.14 (Pearson, 1990), HFCS owes its profitability only to sugar price protection. Between 1975 and 1986, the share of HFCS in the consumption of caloric sweeteners thus increased from 4.2 per cent to 33.9 per cent.

The case of Japan is similar to that of the US. Sugar market protection has led to imports being replaced by HFCS, although in Japan, HFCS production is based on imported maize. The EC situation is different, as the community is a net exporter of sugar and has a commitment to import 1.3 billion tonnes of sugar from ACP (African, Caribbean, Pacific) countries at EC support prices. To protect domestic sugar producers, a quota has been introduced which limits the diffusion of HFCS.

The effect of this biotechnological option in combination with restrictive trade policies has been substantial losses of income and employment and drastic adjustment problems in developing countries exporting sugar. In the Philippines, for example, sugar production fell from 3.5 million tonnes in 1980 to 1.8 million tonnes in 1987, inducing serious job losses.

However, the discussion of HFCS impacts tended to overlook the fact that developing countries importing sugar were likely to benefit from HFCS and from the increased price competition which it brought into the sugar market. In 1987, developing countries exported about 17 million tonnes of sugar which were perhaps threatened by HFCS and protectionism; but other countries, and among them some of the poorest, had to import 14 million tonnes of sugar (Heidhues and von Oppen, 1990). Discussing only the sugar exporting position of the Third World does not do justice to these differences, nor to the potential gains for the sugar importers.

The HFCS example shows that substitution can be fast and substantial if there is an economic incentive, although in this case, the incentive was a side-effect of agricultural protectionism. It also shows that impacts on developing countries can vary considerably, according to the particular conditions. However, although the success of HFCS was linked to protectionist policies, it would be a mistake to ignore the fact that other new enzyme technologies with broad applications will contain a large potential for technological substitution that, in the long term, could create surprises and adjustment problems.

d)   Access to biotechnology

Access to and competence in biotechnology are another crucial factor shaping the impacts on competitiveness and trade. The conditions of R&D in agrofood biotechnology differ substantially from those prevailing during the "green revolution". The latter was based on public R&D in international agricultural research centres (IARCs) and their counterparts, the national agricultural research centers (NARCs) in the developing world. In contrast, biotechnologies depend primarily on private investment and are protected by intellectual property rights, for which adequate economic, legal and institutional bases are lacking in many developing countries. The amounts of investment in agrobiotechnologies indicate the imbalance. Industry investment in the United States was estimated at $300 million in 1987 compared to $85 million spent by the US Department of Agriculture. The Consultative Group on International Agricultural Research (CGIAR) – the major international publicly financed crop research organisation – spent only $11.6 million on biotechnology research in 1989. Even substantial increases in CGIAR investment now would yield results for farmers only after a period of at least ten years. Only about ten developing countries have major biotechnology research programmes, including Brazil, Mexico, China, India, Indonesia, Malaysia, Singapore and Thailand. The poorer developing countries, e.g. in Africa, do not belong to this group. Hence, in the absence of policy measures, divergences are expected to grow between these two groups of developing countries.

Two conclusions can be drawn: first, the impacts of biotechnologies on developing countries as a whole will not be as dramatic as was believed a few years ago. Second, detailed technology assessments and specific policies are required, addressed particularly to the countries with no R&D capacity in biotechnology, those which are negatively affected, or those which need help to solve their hunger problems (Persley, 1990).

6.   Impacts on Employment

Biotechnologies could exert an influence on agricultural employment by two routes: a direct one by application in agriculture and an indirect one by application in upstream and downstream sectors. In both cases the impacts could be job-destroying or job-creating. Furthermore, results will be different for closed or for open economies and particularly for well protected or less protected agricultural sectors. The impacts on structure, on international trade and competitiveness and on employment are obviously closely linked (OECD, 1989).

Total employment in agriculture shows the employment potential which might be affected. As has been pointed out, the share of agriculture in total employment is declining in OECD countries. The OECD total fell from 10.6 per cent in 1978 to 8.9 per cent in 1985. In highly industrialised OECD countries with large farms such as the USA and the United Kingdom, the share declined to 3.1 and 2.5 per cent in 1985, respectively. Mediterranean countries reach proportions of 23.1 and 23.9 per cent for Portugal and Greece respectively and 57.1 per cent for Turkey. As the economic, political and technical forces behind this secular process are still at work, further declines in agricultural employment can be expected. However, for the first group of countries, the potential for further decline is now quite limited, whereas the latter group has a substantial potential for further employment reductions and has in fact, experienced high rates of decline in the recent past. Biotechnology, as one major source of technical change among others, will contribute to this trend.

However, even in highly industrialised countries with low rates of agricultural employment, concern about expected negative employment effects of agricultural bio-technologies has stimulated the rise of countervailing powers to limit their application. Much opposition against the sugar substitute isoglucose, in Europe, and against bST, in various countries, has been the result of such employment concerns.

However, the central questions for the future go widely beyond biotechnology. Will agricultural protectionism be reduced, and if so, at what pace and to what extent? If agricultural protection falls, employment in agriculture will decline, but employment in other sectors will increase due to the "welfare effects" of larger resources being made available elsewhere. Depending on the policy path chosen, the overall effects could be much larger than the impact of biotechnologies alone. For the adjustment of the farming sector, however, the combined effect of all influences is relevant. Policies therefore should take into account effects from both biotechnology and from a reduction of agricultural protectionism.

It has been argued that biotechnology might help to develop new non-food products based on agriculture and thus create new jobs in agriculture. However, assessments in Germany indicate that the potential short-term effects will be small (Meinhold, 1987), principally because of the still insufficient profitability of renewable resources in general.

Depending on the price policy pursued, rates of annual labour reduction in agricul-ture of between 1.5 per cent, for an income-oriented price policy, and 5.5 per cent, for a market-oriented price policy, have been projected for the period 1986 to 2000. In Germany, agricultural labour units would decline from 560 000 to 460 000 in the first scenario and to 234 000 in the second scenario. Production of new renewable resources is projected to employ not more than approximately 13 000 labour units in agriculture, and only with government support. Unless this production uses idle land, it will merely replace food products and hence, at best, reduce the rate of decline in the agricultural labour force. It must be stressed that these are short-term estimates. In the longer term, a growing conversion of agriculture to non-food products could lead to additional employ-ment opportunities, but this cannot relieve current pressures on agriculture.

Use of biotechnologies in upstream and downstream sectors will directly influence employment in these sectors and indirectly impinge on employment in agriculture. New jobs in the downstream sectors might be created by quality improvements in food and by non-food uses of agricultural products. In large-scale, capital-intensive processing units the employment effects will be minor. However, if biotechnologies could be used for the development of new high-value food niches and decentralised renewable resource pro-duction ("agro-refineries"), the employment effects are likely to be more substantial. Direct employment effects in upstream sectors are closely linked to the demand for agricultural food and non-food products; in addition, employment shifts can be expected between agrochemical and biotechnology sectors and within companies. The indirect employment effects in agriculture due to biotechnology induced changes in upstream and downstream sectors cannot be evaluated at this time.

Countries with open economies, less protected agricultural sectors, and a large potential for development of agricultural biotechnologies, will win international market shares and hence protect agricultural employment by scientific and market forces, but this will come partially at the expense of slow adopters and with no guarantee of a globally positive net effect. However, as in the case of renewable resources, this effect will in general merely reduce the rate of long-term decline of agricultural employment.

7. A Case Study: Economic Impacts of bST

This section presents a summary of economic impact assessments of bST, subsequent to the scientific assessment provided in Chapter II and to the comments on how industry views bST (Chapter V).

The reason why bST has been chosen is not because it is typical of agro-biotechnology, for it probably is not, nor that it is more important than other agro-biotechnologies. On the contrary, other innovations of agrofood biotechnology could have more lasting and important effects. The reason is simply that bST is the only biotechnological innovation in the agrofood sector that has led to a number of quantitative economic impact assessments focusing on a single product. There are already several dozen studies on the economics of bST, many of them, however, only on particular aspects. The following analysis is based on approximately ten general economic assessments carried out in Europe and the United States.

There are several reasons why economic impacts of bST have been studied more intensely than the effects of any other single biotechnology:

i) bST is the first genetic engineering product in agriculture for which physical data are available on which quantifications for economic impact assessments can be based;

ii) milk and dairy products are a very important part of the agricultural sector and account for one-quarter of agricultural sales in some OECD countries;

iii) the positive product image of milk for human health;

iv) increased public concern about animal production conditions in general;

v) the costs of milk market policies and of milk surpluses in major OECD countries;

vi) the magnitude of the originally alleged yield increases which appeared to exceed impact of other biotechnologies by far, at least in the short run.

It is important to place bST in a more comprehensive technological perspective and thus to understand the probable limits to its relevance even in the milk and dairy context. In fact, it is only one of numerous potential biotechnologies which could influence milk production and processing in the future. It first arrived on the market as a result of rapid scientific progress in one sector, not as a response to specific market needs.

Figure 4 gives a survey of the biotechnologies which are potentially relevant to milk production and processing. At every step, a direct influence of some biotechnologies is conceivable, from breeding of cows to quality control of the final consumer goods. Moreover, biotechnologies might change the quality and profitability of substitutes to current inputs into the milk and diary chain as well, thereby exerting an indirect influence (see left-hand side of Figure 4). The complexity of forward assessments is shown by the figure. Thus, studies on the economic impacts of bST in general are partial because they neglect possible effects of other innovations.

The following analysis proceeds stepwise, from impacts at the farm level, structural changes in dairy farming, and effects at sectoral levels, to consumer and policy responses and to some questions and conclusions.

Figure 4. **Biotechnology impacts on the milk and dairy chain**

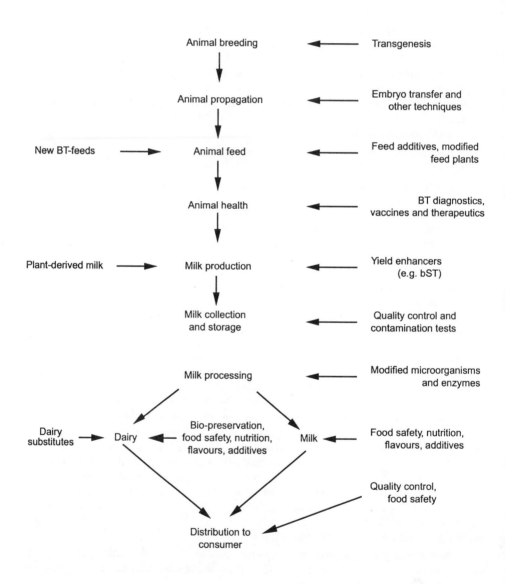

Substitution by New Biotechnologies:
Indirect Impacts

New biotechnologies:
Direct Impacts

Animal breeding ◄——— Transgenesis

Animal propagation ◄——— Embryo transfer and other techniques

New BT-feeds ———► Animal feed ◄——— Feed additives, modified feed plants

Animal health ◄——— BT diagnostics, vaccines and therapeutics

Plant-derived milk ———► Milk production ◄——— Yield enhancers (e.g. bST)

Milk collection and storage ◄——— Quality control and contamination tests

Milk processing ◄——— Modified microorganisms and enzymes

Dairy substitutes ———► Dairy ◄——— Bio-preservation, food safety, nutrition, flavours, additives ———► Milk ◄——— Food safety, nutrition, flavours, additives

Quality control, food safety

Distribution to consumer

a) *Impacts at the farm level*

The effects of bST at the farm level depend on several sensitive but uncertain parameters: the production response of milk cows treated with bST, the bST price, the return on expenses required by farmers to adopt bST, and the milk price after introduction of bST. Moreover, political and economic changes in the future would have to be anticipated for assessments of bST impacts. In most studies these parameters have been fixed by assumptions, in others sensitivity tests have been made, including different political scenarios. As studies for different regions and countries have been based on different assumptions, and as natural and economic conditions for milk production are specific to geographic location, the results can be compared only with the greatest caution.

Yield increases under normal farm conditions will be lower than they have been during trials. Daily milk increases of 17 to 25 per cent have recently been rated as realistic during the period of treatment. On an annual basis this amounts to milk yield increases of between 11 and 16 per cent. A new survey indicates that the expected yield increases will not be proportional to original yield levels, but will be the same at all yield levels for cows producing 5 500 to 9 000 liters of milk per year (OTA, 1991). In the first case the economic incentive for the use of bST would be bigger for high-yielding herds, in the latter case for cows with low yields. However, herd management is a critical factor for efficient use of bST, and better herd management will result in higher yields; thus, profitability is expected to increase with the level of yields.

Costs of bST are still not known in 1991. Most studies calculate gross increases in profits, from which bST costs have to be deduced.

The impacts of bST at the farm level, assuming no bST influence on the price of and demand for milk, and a bST price of zero, can be summarised in the following trends: a given quantity of milk can be produced with fewer cows, but feeding ratios have to be adjusted to the higher yield level. Beef and veal production will decrease, and the demand for labour and land as well as for building capacity used in milk production will be reduced. Constant milk sales, decreasing costs for feed, and decreasing sales of beef, veal or calves will leave only a small surplus due to the use of bST that will hardly cover the costs for bST and its application.

The overall profitability of bST use, aside from special applications, depends crucially on the opportunity costs for labour, land and building capacity. In a policy scenario with no change in milk support prices, and with no quantitative limits for milk producers, these production factors will be used for the expansion of milk supply, leading to increasing self-sufficiency or to surpluses. Under a milk quota regime, building capacities and land may be used for specialised beef, sheep or other livestock production, and land for the production of hay, silage or grain for sale. Labour will be reallocated accordingly, in special cases even to off-farm employment. Under market conditions, however, a mixture of milk production expansion and factor-diverting strategies for other production can be expected.

In the Netherlands, a country with high opportunity costs for the factors mentioned, bST-induced increases of labour income have been estimated to be in the range of Gld 160-300 (i.e. $48-90) per cow, based on the assumption of a 20 per cent increase in milk yields and without taking into account bST costs (Giesen, Oskam and Berensten, 1989). The herd size of the model farms which have been studied was in the range of 40-70 cows. The differences in profitability depend largely on differences in opportunity costs for land, labour and building and partly on differences in initial yield between the farm groups analysed. Farms with more than 2.35 cows per hectare could improve labour income by 25 per cent, farms with less than 1.9 cows per hectare gained 20 per cent.

Comparisons of studies for five European countries reveal wide differences: for many German farms the use of bST would not be profitable, although for special groups results similar to those for the Netherlands have been calculated. On the other hand, even higher rates of profitability than in the Dutch analysis have been estimated for large producers in Italy and the UK. Similar levels and variations in expected profits have been found in US studies.

b) Impacts on structural change

At first sight, bST appears to be a scale-neutral technology because it does not require great amounts of capital. Nevertheless, an indirect scale effect is expected, with successful use of bST depending on management quality. One indicator of management quality is milk yield, which tends to increase with herd size, despite variations of milk yield within the same herd size class. Hence, higher adoption rates of bST are expected for larger herds. Moreover, if bST has to be applied by a veterinarian, fixed costs for a farm visit would add another incentive for higher adoption rates in larger herds.

As the farm level analysis indicated, bST use is profitable only under specific farm conditions. As the response in yields varies substantially from cow to cow and as further uncertainties are connected with the introduction of bST, farmers who adopt it are expected to use it only on part of their herd.

The structural impact of bST has been a major concern in public debate, considering that yield increases in the range of 25-40 per cent have been forecast. However, these rates are widely exaggerated. They must be deflated for a number of reasons:

  i) increases of 25-40 per cent are effects which have indeed been obtained for single cows, but the variations in yield responses are very large; some cows do not respond, and it is not even known if there have been cases of negative response; hence, the average yield increase of a herd is substantially lower than the maximum individual response;

  ii) as bST is a management-intensive production input, yield increases at the farm level can be expected to be lower than the increases obtained during scientific trials;

  iii) bST use will not be profitable under all farm conditions in the short run, and not for all cows of a herd, which means that the maximum adoption rate will be significantly below 100 per cent.

Therefore, recent studies expect a more gradual and less extensive adoption of bST than early assessments. On the other hand, the bST-induced impact on structural change is a partial effect which comes on top of the normal, quite substantial rate of change. Other technological advances, e.g. in breeding or in milking robotics, might add other partial effects, promoting structural change.

In a study covering nine EC countries, structural changes in milk production without bST were estimated to show the following trends for the period 1985 to 1995 (Buckwell, 1987): the number of producers will fall by 37 per cent; average herd size will increase by 33 per cent; the total number of cows will decrease by 17 per cent and milk yields will rise by 13 per cent. These estimates assume continuation of the trends that determined structural changes before the quota system was introduced. Reduction of farms and specialisation in production have been the main features in this development.

However, if bST is introduced, an additional yield increase of 15 per cent and an adoption rate of 7 per cent of producers in 1995, controlling 30 per cent of the cows, has been assumed for all nine countries. As a result, 4.4 per cent of the cows can be taken out of production, and the average yield could rise by 4.6 per cent. If the bST adopters were able to use bST on all their cows and extend milk production through bST yield increases by purchasing quota from non-adopters, the number of small producers would fall by 8 per cent.

In the long run, i.e. after 1995, structural effects will increase, because bST accelerates structural change in milk production, and structural change in milk production increases the ratio of potential adopters.

In any event, the main changes brought about by bST will be long term, and thus, there will be time to anticipate them and prepare for them.

c) *Impacts at national level*

Quantitative changes at farm level can lead to changes in sectoral factor demand and product supply and to corresponding changes in prices, government expenditures and consumer welfare. As impacts depend largely on policy scenarios, only conditional results can be summarised here.

For the milk market two extreme scenarios of a US study can be mentioned (Fallert, 1987): a strategy of constant prices in order to help small farmers to survive, and a policy of milk price adjustments to achieve an equilibrium in the milk market. In the first case, government intervention and expenditure will increase substantially; in the second, a given yield increase due to bST will lead to a 9 per cent fall in milk prices. Recent studies emphasised the sensitivity of the results to various factors and estimated that the effects of bST will be lower (Marion and Wills, 1990; Tauer and Kaiser, 1991).

In a quota system the primary effect of bST will be cost reduction. It is likely that agricultural price policy will take these cost cuts into account when fixing the following year's price support line. If price policy is oriented towards average production costs, bST adopters would still enjoy a bST rent, although a reduced one; non-adopters, however, would experience price pressure. In a quota system the major route to adjustment at the farm level is diversion of factors to other types of production. Specialised beef production will take advantage of idle labour, land and building capacities, and will also profit from price incentives for beef resulting from the reduction of milk cow numbers. In the Netherlands, an increase of 63 per cent of beef bulls has been estimated as a result of bST (Giesen, Oskam and Berentsen, 1989).

This result sheds new light on two other arguments in the bST discussion: the use of bST has been proposed as a means to reduce manure pollution and possible greenhouse effects due to cattle, as the same amount of milk could be produced by fewer cows with higher yields. However, this environmental effect has to be calculated against the consequences of the induced increase in beef bulls and sheep, before a conclusion can be drawn on the net effect.

Further impacts on the pattern of production have been calculated. One impact could be an increase in maize acreage. Shifts of free factors to non-milk production might transfer market pressures to other markets. This in turn will lead to higher intervention or lower prices and respective distributional effects for the various groups of producers, consumers and taxpayers. Total increase in annual national income for the Netherlands, assuming a 20 per cent yield increase and adoption for 28 per cent of all cows, was estimated as $36 million, from which the costs of bST would have to be deducted.

d)   Consumer response

Chapter VI reviewed the general issue of public and consumer response to biotech-nology, which is of crucial importance to the future of bST. However, most economic studies of bST tend to ignore this problem because its assessment is so difficult. Opinion polls of farmers, dairy processors, or consumers that attempt to measure acceptance of bST have produced mixed results in some cases and negative ones in others. Although quality and food safety can, to some extent, be defined according to purely scientific criteria, public perception follows other criteria as well. Thus, the public increasingly demands information not only on product content, but also on production processes. It is an open question whether ongoing scientific investigations, such as those undertaken in the EC, will change public attitudes significantly. If bST were approved and introduced into the market, some of the skeptical farmers could be expected to change their minds about bST under the pressures of competition. However, it is the consumer who will remain the major critical decision factor.

Two sub-groups have to be distinguished. The first comprises people who are indifferent to bST and who buy in response to economic incentives only (Beusmann, 1989). These consumers could gain a consumer rent, if bST-induced cost reductions are transmitted to the final consumer. However, as the farm-gate milk price is approximately half the consumer price of milk products and as the elasticity of consumer demand in response to price cuts for milk products is low in industrialised countries (about 0.3), this effect will be minor.

The second sub-group are the opponents. The questions are how long their aversion will last and what their reaction will be. If bST is approved, some consumer groups are likely to request labelling, indicating use or non-use of bST in milk production in order to assure "consumer sovereignty". Opinion polls among dairy processors and consumer organisations in Germany have shown that sensitivity of consumers is expected to be higher for fresh milk products than for processed milk products. Consumer groups would recommend consumption of milk produced without bST or of milk produced on farms managed according to regulations of the International Movement of Organic Agriculture.

The latter category of farms covers at the present time less than 1 per cent of production capacities but is increasing rapidly due to economic pressures on conventional farms and EC subsidies for farm conversion under the extensification guideline. Super-market chains that use the environmental debate to improve their image and business might be part of the forces developing this second line.

However, the creation of two lines of milk and milk products would increase the costs for collection, processing and distribution. The net effect of such a market division on the quantities and prices of both lines is highly uncertain. Another scenario would be a temporary decline in the demand for milk (OTA, 1991).

e)   Conclusions and issues

There is no doubt that bST has attracted more public and scientific attention than other agrobiotechnologies largely because of the apparent size of its solitary yield effect. Other biotechnologies, such as the new breeding techniques, have smaller annual yield effects, but those can continue and accumulate over a longer period, resulting in larger end-effects. Thus, while bST might be typical of some yield-enhancing technologies, it is certainly not typical of agrofood, or even animal biotechnology in general.

A number of economic conclusions can be drawn:

i) Over a longer period, bST could have a significant influence on milk production, farm structures, agricultural markets and milk consumption.

ii) However, the size of the direct economic impacts on dairy farming and milk production is presently expected to be smaller, and the time period necessary for adoption longer, than was originally assumed. Early estimates of milk yield increases of 25 to 40 per cent have given way to estimates in the range of 10 to 16 per cent, and the time for adoption and diffusion is expected to last from five to ten years. If bST turns into a major innovation, it will be one among others and will increase productivity in agriculture gradually rather than abruptly.

iii) The economics of bST will be crucial. Present indications, variable according to country and place, are sometimes positive, sometimes not. Public acceptance of bST and the associated government policies remain the single most critical factor, with a potential for substantial repercussions on the economics of milk production. In the future more efforts will have to be made to anticipate which biotechnologies might face acceptance problems and for what reasons.

iv) In the case of acceptance problems, two basic decision routes are possible: a political or a market decision. The EC decision prohibiting steroid hormones is an example of the first case. Negative trade impacts might be avoided in such situations by permitting trade of goods manufactured in various countries under the same standards. In contrast, a market decision would evolve if, after approval of a biotechnology product, the final decision is left to producers and consumers. For both the political and market routes, the consequences of a decision can be evaluated before it is made, and public participation in decision-making can be assured.

8.   Summary

Economic *ex ante* assessments of expected biotechnology impacts face numerous uncertainties due to unpredictable political developments, technological complexity and methodological assessment problems. A reduction of agricultural protectionism will slow down biotechnological innovation and diffusion in high-price countries or sectors, but after a period of structural change and adjustment greater efficiency will lead again to faster adoption of innovation. Also, cost-reducing biotechnologies could help alleviate agricultural adjustment problems.

Aggregate estimates of impacts on agricultural production and market balances of the US and the EC reveal a big potential of biotechnologies to increase productivity by decreasing costs per unit of output or by increasing yields per unit of input. These effects are likely to be partly offset by the expected decline of general productivity increases linked to traditional technological factors.

The only new biotechnology product for which a number of precise economic impact assessments exist is bST. They show that, even for a product with a big productivity-enhancing capability in a large sector of agriculture, the main direct result (in a quota system) might be cost reduction, and that production increases might take more time than originally expected, come gradually and be smaller than first individual trials have suggested.

Biotechnology, like other technical changes, could lead to downward pressure on food prices and thus to structural changes at the farm and employment levels. While biotechnology could play a leading role in the development of non-food products from agriculture, conversion towards such products is hampered by policies supporting farm-gate prices for food, and by the fact that the price of non-renewable resources does not sufficiently internalise long-term scarcities.

Biotechnology will accelerate the ongoing structural changes of the agrofood sector towards increased vertical integration of agriculture with upstream and downstream sectors. Biotechnology adoption rates are expected to increase with management quality, which means that they are likely to increase with farm size (measured by income potential, not acreage). For the time being, it is not possible to estimate the consequences of this concentration process on agricultural input prices. It might be necessary to monitor this process and to strengthen the bargaining position of agriculture with regard to the upstream and downstream sectors.

International trade and competitive advantages might shift towards countries which have both good capabilities in biotechnology and viable farms, if trade policies allow for such shifts. Although the negative impacts of such shifts on the Third World in general have probably been overestimated, their consequences for some developing countries must be monitored and policies devised to help the latter to develop and use biotechnologies.

Biotechnology is expected to contribute to the ongoing long-term trends reducing the share of total employment in agriculture. International leaders in biotechnology might increase their international market shares and, thus, save employment in agriculture, but this would come at the expense of the lagging countries with no guarantee of a positive net effect on employment.

Biotechnologies are related or relevant to nearly all major agricultural policy questions of our time. When they lie at the very heart of a problem, as is often the case, they may provide new and better solutions (reduction of chemicals, of factor costs, better food safety, food for the Third World) or they may tend to aggravate it (market imbalances, unemployment, concentration), or they may do both at the same time.

As biotechnologies will increasingly become central to technical change in agriculture and food production, agricultural policies will have to cope with them on a permanent basis, by following, selecting and using them. The main message is that biotechnology in agriculture is not a global issue, but one that must be addressed in a differentiated way, case by case.

9. Further Reading

ALEXANDRATOS, N., ed. (1988), *World Agriculture: Towards 2000, An FAO Study*, Food and Agriculture Organisation, Belhaven Press, London.

ANTONELLI, G. and A. QUADRIO-CURZIO, eds. (1988), *The Agro-Technological System Towards 2000: A European Pespective*, Amsterdam, Elsevier Science Publishers BV (North-Holland).

BARKER, R. (1990), "Socio-economic Impact", in *Agricultural Biotechnology: Opportunities for International Development*, G.J. Persley, ed., CAB International, Wallingford, pp. 299-310.

BEUSMANN, V. *et al.* (1989), "Folgen des Einsatzes von BST in der deutschen Milcherzeugung", *Schriftenreihe des Bundesministers für Ernährung, Landwirtschaft und Forsten, Reihe A: Angewandte Wissenschaft*, H 346, Landwirtschaftsverlag, Münster-Hiltrup.

BONNEY, S. and P. DAUCE (1989), "Les nouvelles technologies en agriculture. Une approche technique et économique", *Cahiers d'économie et sociologie rurales*, No. 13, pp. 5-34.

BUCKWELL, A. (1987), "Structural Impacts of Bovine Somatotropin in the Dairy Sector of the European Community" (unpublished report), Wye College, University of London.

BUCKWELL, A. and A. MOXEY (1990), "Biotechnology and Agriculture", *Food Policy*, Vol. 15, pp. 44-56.

COCHRANE, W.W. (1958), *Farm Prices – Myth and Reality*, University of Minnesota Press, Minneapolis.

COHEN, J.I. (1990), "International Donor Support for Agricultural Biotechnology", *Food Policy*, pp. 57-65.

DE HAEN, H. and F. ISERMEYER (1989), "Konsequenzen gentechnischer Entwicklungen für die Landwirtschaft", in *Landwirtschaftlicher Strukturwandel in den 90er Jahren*, ed. J. Jarre, Loccumer Protokolle 5/1989, Evangelische Akademic Loccum.

DIXIT, P.M., M.R. HERLIHY and S.L. MAGIERA (1989), "Global Implications of Agricultural Trade Liberalisation", in *Agricultural Food Policy Review: US Agricultural Policies in a Changing World*, US Department of Agriculture, Economic Research Service, AER 620, Washington, DC, pp. 253-274.

Enquete-Kommission (1989), "Gestaltung der technischen Entwicklung; Technikfolgen-Abschätzung und-Bewertung des Deutschen Bundestages, Zum gentechnologisch herges-tellten Rinderwachstumshormon", Deutscher Bundestag, 11, Wahlperiode, Drucksache 11/4607.

FALLERT, R. *et al.* (1987), "bST and the Dairy Industry", *A National, Regional and Farm-level Analysis*, USDA/ERS, Agricultural Economic Report No. 579, Washington, DC.

FLEISCHER, B. (1989), "The Economic Risks of Deliberately Released Genetically Engineered Microorganisms", *American Journal of Agricultural Economics*, Vol. 71, pp. 480-488.

GIESEN, G.W.J., A.J. OSKAM and P.B.M. BERENTSEN (1989), "Expected Economic Effects of BST in the Netherlands", *Agricultural Economics*, Vol. 3, pp. 231-248.

GOODMANN, D., B. SORJ and J. WILKINSON (1987), *From Farming to Biotechnology – A Theory of Agro-Industrial Development*, Basil Blackwell, Oxford, UK.

GOTSCH, N.A. (1990), "Zukünftige biologisch-technische Entwicklung im Pflanzenbau", *Ergebnisse einer Delphi-Studie und Modellrechnungen unter verschiedenen Szenarien*, Diss, ETH Zürich No. 9193.

HEIDHUES, F. and M. VON OPPEN (1990), "Auswirkungen biotechnischer Neuerungen auf die Landwirtschaft in Entwicklungsländern", in *Schriften der Gesellschaft für Wirtschafts – und*

Sozialwissenschaften des Landbaues eV, eds. H.E. Buchnolz, E. Neander and H. Schrader, Bd 26, Landwirtschaftsverlag, Münster-Hiltrup.

HOFREITHER, M.F. (1990), "Was bringt die Liberalisierung des Welthandels?", *Agrarische Rundschau,* 4, pp. 26-29.

ISERMEYER, F. and H. DE HAEN (1989), "Rinderwachstumshormon rbST - Viele Fragen offen", Tierzüchter.

JOLY, P-B. (1989), *Trajectoire technologique et stratégie des agents de l'industrie des semences,* INRA, Actes et Communications, No. 4, Grenoble, France.

JUNNE, G. (1990), "The Impact of Biotechnology on International Commodity Trade", mimeo, University of Amsterdam.

KALTER, R.J. and L.W. TAUER (1987), "Potential Economic Impacts of Agricultural Biotechnology", *American Journal of Agricultural Economics,* Vol. 69, No. 2, pp. 420-425.

KENNEY, M. (1986), *Biotechnology – The University-Industrial Complex,* Yale University Press, New Haven and London.

KLOPPENBURG, J.R. Jr. (1988), *First the Seed, The Political Economy of Plant Biotechnology, 1492-2000,* Cambridge University Press, Cambridge.

KNERR, B. (1990), *Biotechnology and the World Sugar Market,* European Association of Agricultural Economists, VIth Congress, The Hague.

KUCHLER, F. and S. OFFUTT (1986), "US Congress, Office of Technology Assessment, Technology, Public Policy and the Changing Structure of American Agriculture", Book review, *American Journal of Agricultural Economics,* Vol. 68, pp. 764-766.

LIPTON, M. and R. LONGHURST (1988), *New Seeds and Poor People,* London.

LONGWORTH, J.W. (1987), "Biotechnology: Scientific Potential and Socio-economic Implications for Agriculture", *Review of Marketing and Agricultural Economics,* Vol. 55, pp. 187-199.

MARION, B.W. and R.L. WILLS (1990), "A Prospective Assessment of the Impacts of Bovine Somatotropin: A Case Study of Wisconsin", *American Journal of Agricultural Economics,* Vol. 72, pp. 326-336.

MEINHOLD, K. *et al.* (1987), *Möglichkeiten und Grenzen beim Anbau regenerativer Rohstoffe für Energieerzeugung und chemische Industrie.* Studie im Auftrag der Enquete-Kommission Technologiefolgenabschätzung des Deutschen Bundestages. Materialien zu Drucksache 10/6801, Band IV, Bonn.

MELLOR, J. (1990), "Global Food Supply: Sustainable Agriculture and Natural Resource Base", in *NAVF (Norwegian Research Council for Science and the Humanities), Sustainable Development, Science and Policy,* Oslo.

MÜHLEBACH, F. and U. BEERNEGGER (1990), "Wie beeinflussen Embryotransfer, Geschlechtsbestimmung und Leistungsförderer die Schweizer Landwirtschaft?", in *Technischer Fortschritt in der Landwirtschaft, Schriften der Gesellschaft für Wirtschafts – une Sozialswissenschaften des Landbaues eV,* Vol. 26, eds. H.E. Buchholz, E. Neander and H. Schrader, Landwirtschaftsverlag, Münster-Hiltrup.

NEVILLE-ROLFE *et al.* (1988), "Report on the Impact of Biotechnology on Agriculture in the EC to the Year 2005", mimeo prepared by the Bureau Européen de Recherches SA.

OECD (1988), *Biotechnology and the Changing Role of Government,* Paris.

OECD (1989), *Biotechnology: Economic and Wider Impacts,* Paris.

OECD (1990), *Modelling the Effects of Agricultural Policies,* OECD Economics Studies, Special Issue, No. 13, Paris.

OTA (1988), *Technology, Public Policy and the Changing Structure of American Agriculture,* OTA-F-285, US Congress, Washington, DC.

OTA (1991), *The US Dairy Industry at a Crossroad: Biotechnology and Policy Choices,* Special Report, OTA-F-470, US Government Printing Office, Washington, DC.

PAULINO, L. (1986), "Food in the Third World: Past Trends and Projections to 2000", Research Report 52, IFPRI.

PEARSON, S. R. (1990), "Substitution in End Uses for Food Commodities and Agricultural Trade Policy", *Food Research Institute Studies,* Vol. 22, pp. 109-127.

PERSLEY, G. (1990), *Beyond Mendel's Garden: Biotechnology in the Service of World Agriculture,* The World Bank, CAB International, Wallingford, Berkshire, UK.

RUTTEN, H. (1989), *Technical Change in Agriculture: A Review of Economic Literature, with Special Reference to the Role of Prices,* Agricultural Economics Research Institute, The Hague.

STRANGE, M. (1988), *Family Farming - A New Economic Vision,* University of Nebraska Press, Lincoln and London.

TAUER, L.W. and H.M. KAISER (1991), "Optimal Dairy Policy with Bovine Somatotropin", *Review of Agricultural Economics,* Vol. 13, pp. 1-17.

THIEDE, G. (1987), "Milcherzeugung im Somatotropin-Zeitalter", *Agra Europe* (German edition), No. 37, Dokumentation.

THOMPSON, P. B. (1990), "Agricultural Ethics and Economics", *The Journal of Agricultural Economics Research,* Vol. 42, No. 1, pp. 3-7.

TWEETEN, L. and M. WELSH (1987), "The Economics of Agricultural Biotechnology: Discussion", *American Journal of Agricultural Economics,* Vol. 69, pp. 440-442.

YOUNG, E. (1990), "Bread for the Hungry?", *World Watch,* pp. 14-22.

LIST OF PARTICIPANTS

The following experts participated in the OECD Expert Seminars at Novara, Italy (6-8 March 1990) and Château de Bellinglise, France (15-17 May 1991) and/or made written contributions to the text of this publication.

CHAIRMAN

Prof. Daniel THOMAS
Head "Programme mobilisateur
    biotechnologies"
University of Compiègne,
Compiègne
France

Dr. Volker BEUSMANN
Institute of Farm Economics
Research Establishment for Agriculture
Braunschweig
Germany

Dr. Robert COLLIER
Manager of Animal Sciences,
Corporate R&D
Monsanto Company
Missouri
United States

R. Stephen CRESPI
Patent Consultant
Bognor Regis
Sussex
United Kingdom

Prof. Sir Barry CROSS
Honorary Secretary of the Zoological
Society of London
President of Corpus Christi
College, Cambridge
United Kingdom

Prof. Richard B. FLAVELL
AFRC Institute of Plant
Science Research, and
Director, John Innes Institute
Norwich
United Kingdom

Dr. Riccardo GALLI
Professor of Technology Assessment
Milan
Italy

Prof. Douglas L. GEORGALA
Director
AFRC Institute of Food Research
Shinfield, Reading
United Kingdom

Dr. Christina GOODACRE-FORSLING
Programme Manager
Food Processing Sciences
Laboratory of the Government Chemist
Teddington
United Kingdom

Dr. Willy De GREEF
Consultant on Plant Biotechnology
Ghent
Belgium

Prof. R. Brian HEAP
Director of Research
Institute of Animal Physiology
  & Genetics Research
Babraham Hall
Cambridge
United Kingdom

Dr. Ernest JAWORSKI
Distinguished Science Fellow
Monsanto Company
St. Louis, Missouri
United States

Prof. Heinrich KARG
Institute for Animal Sciences
Technical University
Munich
Germany

Prof. Louis LEMKOW
Professor of Sociology
Autonomous University
Barcelona
Spain

Dr. Etienne MAGNIEN
Commission of the European
Communities, DG XII
Biotechnology Division
Brussels
Belgium

Dr. Nobuo MURATA
Director
Eco-Physiology Research Division
Tropical Agriculture Research Center
Tsukuba-City
Japan

Dr. Vincent PETIARD
Director
Plant Biotechnology Research
Francereco S.A. Nestlé
Notre Dame D'Oé
France

Dr. Brian RICHARDS
Chairman
British Bio-technology Ltd.
Oxford
United Kingdom

Prof. R.J. ROBERTS
Director
Institute of Aquaculture
University of Stirling
Scotland
United Kingdom

Prof. Dr. R.A. SCHILPEROORT
Professor of Plant Molecular Biology
Leiden
Netherlands

Dr. Albert SIMANTOV
President,
Centre international de hautes études
   agronomiques méditerranéennes
Paris
France

Dr. Bernardo SORJ
Director, Sorj & Wilkinson
Consultants
Rio de Janeiro
Brazil

Prof. Dr. Gerald STRANZINGER
Institute for Animal Science
Federal Institute of Technology (ETH)
Zurich
Switzerland

Prof. Sue A. TOLIN
Professor of Plant Pathology
Virginia Polytechnic Institute and
State University
Blacksburg
United States

Dr. John WILKINSON
Postgraduate Centre for Agricultural
Development
Federal University
Rio de Janeiro
Brazil

OECD

Dr. Michael West OBORNE
Head, Science and Technology Policy
  Division

Directorate for Science, Technology
  and Industry (DSTI)

Paris
France

Dr. Salomon WALD
DSTI

Ms. Bruna TESO
DSTI

Mrs. Carliene BRENNER
Development Centre

Dr. Jan DROZD, Consultant
Sittingbourne, Kent
United Kingdom

SPONSORS

Mr. Marc CHOPPLET
Secretary General
Le Biopôle
Amiens
France

Prof. Amilcare COLLINA
General Manager, FERTEC
Co-ordinator of R&D
Ferruzzi Group
Milan
Italy

GLOSSARY

Actin	One of the two contractile proteins in muscle
Aerobic bacteria	Bacteria that can only live in the presence of oxygen
Agar	Polysaccharide gelling agent
AI	Artificial insemination
Alginate	Polysaccharide gelling agent
Alphalactalbumin	Protein component of milk
Amino acids	The building units of proteins; amino acids are linked together in a particular order which determines the character of different proteins
Amylopectin	The water-insoluble portion of starch
Amylose	The water-soluble portion of starch
Anaerobic digestion	Fermentation of effluent materials in the absence of oxygen, often with the generation of methane
Anther	The part of the stamen that produces the pollen
Antigen	A macromolecule (usually a protein or carbohydrate) which, when introduced into the body of a human or higher animal, stimulates the production of an antibody that reacts specifically with it
Antinutrients	Compounds that inhibit normal uptake of nutrients
Antisense genes	Genes in which the mirror image of the normal nucleotide base sequences are inserted, preventing expression of the natural genes
Apical dominance	The phenomenon in which the presence of a growing apical bud on a plant inhibits the growth of lateral buds

Assay	The procedure for measuring the quantity of a given substance in a sample
Biocatalysis	Use of enzymes to catalyse chemical reactions
Biomass	All organic matter that derives from the photosynthetic conversion of solar energy
Biosphere	The part of the earth and its atmosphere that is inhabited by living organisms
Biotope	A small habitat in a large community
Blastocyst	A mammalian egg in the later stages of cleavage. It consists of a hollow ball of cells
Blastomere	The products of cleavage of animal eggs; the egg usually divides into two, then four, then eight blastomeres and so on
Bovine somatotropin	Growth hormone that can be produced by recombinant DNA technology and used to increase the milk yield in cows
Brucellosis	Disease caused by infection with organisms of the genus Brucella
Calf scours	A watery diarrhoea in calves
Callus	An undifferentiated mass of cells that forms on plant wounds and is also produced by plant tissues growing in cell culture
Canine	Affecting dogs
Canola	Vegetable oil high in monounsaturated fatty acid
Casein	A phosphoprotein component of milk
Cellulase	Enzyme catalysing the breakdown of cellulose
Chimera	In animals, an individual exhibiting two or more genotypes in patches derived from two or more embryos
Chitin	A nitrogenous polysaccharide occurring as skeletal material in many invertebrates
Chitinase	An enzyme which breaks down chitin, a component of, for example, cell walls of fungi
Chloroplasts	Cellular organelles where photosynthesis occurs
Chymosin	An enzyme used to clot milk; it is used in the manufacture of cheese

Chromosome	Chemical packages of hereditary information (genes), made up of long coiled chains of DNA. Found in the nucleus of cells
Clone	A collection of genetically identical cells or organisms which have been derived asexually from a common ancestor; all members of the clone have identical genetic composition
Coding	The specification of a peptide sequence, by the code contained in DNA molecules
Cogeneration	Production of both electricity and process heat (steam) in one industrial plant
Cross	Breeding from dissimilar parents or the products of such cross breeding
Cross pollination efficiency	Efficiency of pollen from one plant reaching the stigma of another plant
Cryogenic	At very low temperature
Cultivar	Any agricultural or horticultural variety. The term is derived from cultivated variety
Cyclodextrin	Cyclic polymer of dextrose
Cysteine	An amino acid
Cytoplasm	The living contents of a cell, excluding the nucleus and large vacuoles
Cytoplasmic genes	DNA-containing bodies in the cell but external to the nucleus
Cytoplasmic organelles	Discrete subcellular structures located in the cytoplasm of cells; these allow division of labour within the cell
Diploid	The state of having two complete sets of match-paired chromosomes, one set of parental origin and the other of maternal origin
Disaccharides	Carbohydrate consisting of two linked sugar units
DNA	Deoxyribonucleic acid; polymer composed of deoxyribonucleotide units; genetic material of all organisms except RNA viruses
DNA probes	Isolated single DNA strands used to detect the presence of the complementary (opposite) strands. Can be used as very sensitive biological detectors
Ecosystem	The complex of a community and its environment functioning as an ecological unit in nature

Electrophoresis	Method of separating molecules on the basis of electrical charge
Embryo	An immature organism in the early stages of its development
Embryo transfer	Implantation of embryos from donor animals or generated by in vitro fertilisation into the uteri of recipient animals
Endocrine interference	Interference with the normal balance of hormones
Enucleated ova	Egg cells from which the nucleus has been removed
Enzyme	A protein that catalyses a chemical reaction
Enzyme bioreactor	A reactor in which a chemical conversion reaction is catalysed by an enzyme
Epizootics	Diseases affecting a large number of animals simultaneously
Essential amino acids	Compounds that cannot be synthesised by the body and therefore have to be ingested with feed/ food
Eukaryote	An organism whose cells possess a nucleus
Exotoxin	A toxin released by a bacterium into the medium in which it grows
F1 hybrid	The first generation resulting from a particular cross
Feline	Affecting cats
Flocculant	Agent that causes small particles to aggregate (flocculate)
Fluorescent probe	Probe whose response is based on the fluorescence intensity of individual cells or cell components
Fractionation	Separation
Fructose	Fruit sugar
Furfural	Used industrially as a solvent and as a raw material for synthetic resin
Gamete	A cell capable of fusing with another cell to form a zygote, from which a new individual organism can develop

Gene	The basic unit of heredity: an ordered sequence of nucleotide bases, comprising a segment of DNA. A gene contains the sequence of DNA that encodes one polypeptide chain (via RNA)
Genetic code	The sequence of bases along the DNA molecule; this information provides the code for specific proteins
Genetic engineering	Technologies, including recombinant DNA technologies, used to isolate genes from an organism, manipulate them in the laboratory and inject them into another organism
Genome	The genetic endowment of an organism or individual
Genotype	The genetic make up of an organism
Germination	The beginning of growth in a spore, seed, zygote, etc., especially following a dormant period
Germplasm	The part of an organism that passes its characters on to the next generation. Most of this information is carried by DNA in the chromosomes
Gonad	The reproductive organ of animals that produce sperm or eggs
Halothane	A volatile anaesthetic
Haploid	A cell or organism containing only one representative from each of the pairs of homologous chromosomes found in the normal diploid cell
Helminths	Parasitic worms
Hemicellulose	One of the structural materials in plants, composed of high molecular weight polymers of sugars
Herbicide	Weed-killing agent
Histocompatibility system	The collection of genes coding for antigens present on the surface of nucleated cells; these antigens are responsible for the differences between genetically non-identical individuals that cause rejection of tissue grafts between such individuals
HIV	Human immunodeficiency virus

Homozygous	Having the same two alleles for a given heritable character
Hybrid	Cross between parents of different species
Hybrid seed	Seed produced by crossing genetically dissimilar parents
Hybridisation	1. In molecular biology a procedure in which single stranded DNA or RNA are mixed. If the molecules have similar structures they will hybridise, that is bind to one another. This procedure can be used to give an indication of the degree of relatedness of two genes, or to detect RNAs or DNAs using a specific nucleic acid probe.
	2. The crossing of two species to form hybrids
Hyperthermia	Abnormally high body temperature
Hypothalamic peptides	Peptides generated in the vertebrate forebrain and concerned with regulating the body's physiological state
Immunochemical control	use of immune agents to combat infections
Immunogenicity	The ability to elicit an immune response
Immunosuppression	Artificial suppression of immune response by the use of drugs
Inbred lines	Breeding between closely related individuals
Interferon	A protein that appears in the body during infections by viruses. It is produced by infected cells and inhibits the growth of viruses
Interleukins	A group of proteins that transmit signals between immune cells and are necessary for mounting normal immune responses
Ions	Constituent parts of compounds that dissociate in solution, e.g. ammonium nitrate, potassium sulphate
Lactose	Milk sugar
Laminarin	A storage polysaccharide of the brown algae
Leaf roll	Virus disease, e.g. of potato characterised by curling of leaves
Lepidoptera species	e.g. moths and butterflies
Lignification	Thickening of cell walls through deposition of lignin

Lignin	One of the main structural materials of vascular plants
Lignocellulose	Compounds of lignin and cellulose found in wood, etc.
Liposome	Spherical shell formed when phospholipids are dispersed in aqueous solutions
Luteinising hormone	A pituitary hormone which causes growth of the yellow body of the ovary and also stimulates activity of the interstitial cells of the testis
Lymphokine	Generic name for proteins that are released by lymphocytes to act on other cells involved in the immune response. The term includes interleukins and interferons
Macropropagation	Production of plant clones from growing parts
Malignant	Cancerous
Malting	A process of generating starch-degrading enzymes in grain by allowing it to germinate in a humid atmosphere
Mammary tumours	Tumours of the milk glands
Mastitis	Inflammation of the mammary gland
Meiosis	Nuclear division in the germ cells, which results in the halving of the number of chromosomes so that the egg cells or sperm that are produced receive just one set of chromosomes instead of the double set found in each somatic cell
Meristem	A distinct region of actively dividing cells primarily concerned with growth (e.g. roots and shoots)
Metabolite	Substance that takes part in a metabolic reaction
Metallothionein gene	The metallothionein gene is switched on when certain metals are absorbed by the animal
Microalgal culture	Culture in bioreactors of microalgae; microalgae include seaweeds
Micro-encapsulation	A process of enclosing a substance in very small sealing capsules from which material is released by heat, solution or other means
Microinjection	The technique of introducing very small amounts of material (DNA or RNA molecules, enzymes, cytotoxic agents) into an intact cell through a microscopic needle penetrating the cell membrane

Micropropagation	Use of small pieces of tissue such as meristem grown in culture to produce large numbers of plants
Microtuber	Cultured tissue capable of growing into tuberous plant
Mitochondria	Structures in higher cells that serve as the "powerhouse" for the cell, producing chemical energy
Monoclonal antibodies	Antibodies derived from a single source or clone of cells which recognise only one kind of antigen
Monogastric animals	Animals with simple stomachs that do not ruminate (see ruminant animals)
Monounsaturates	Oils containing monounsaturated fatty acids
Morphology	The structure and form of an organism
Mutant	An organism that has undergone some change as a result of gene mutation
Mutation	Any change that alters the sequence of bases along the DNA, changing the genetic material
Mycorrhizal fungi	Fungi that exist in symbiotic association with roots or other structures of a plant
Myco-protein	Fungal protein
Mycotoxin	Toxic substance of fungal origin
N_2	Nitrogen
Nematodes	A class of worm, often parasitic
Nisin	Antimicrobial compound produced by certain lactic acid bacteria
Nitrate	The only form in which nitrogen can be used directly by plants; a component of chemical fertilizers
Nitrite	Nitrogenous compound associated with fertilizers
Nitrogen fixation	The formation of soluble nitrogenous compounds from atmospheric nitrogen by nitrogen-fixing bacteria in symbiosis with leguminous plants
Non-cariogenic	Having the property of not supporting the development of dental decay (caries)

Nucleic acid	Molecules consisting of a backbone of alternating sugar and phosphate groups from which protrude certain bases; they include DNA, which forms the genetic material, and RNA
Nucleic acid probe	See DNA probes
Nucleotide	Building blocks of nucleic acids
Nucleotide sequence	Sequence of nucleotides in a segment of DNA
Nucleus	The nucleus is a structure within the cell. It is bounded by a membrane and contains the chromosomes
Oestrogen	The generic term for a group of female sex hormones which induce oestrus
Oestrus	In female mammals, the period of sexual excitement and acceptance of the male
Oligosaccharide	carbohydrate consisting of several linked sugar units
Onco-mouse	A mouse that has been genetically modified to incorporate an oncogene; oncogenes cause cells to undergo cancerous transformation
Oocyte	A reproductive cell in the ovary of an animal that gives rise to an ovum
Organellar genes	Genes located on organelles outside the nucleus
Organoleptic	Having an effect on one of the organs of sense, as taste or smell
Ovulation	In mammals, the process of escape of the ovum (egg cell) from the ovary
Par gene	Genes found in bacterial and plant cells involved in partition of plasmids in cells
Parthenogenesis	Development of unfertilized eggs to form new individuals
Parturition	The process of giving birth
Pathogen	Any micro-organism or virus that causes disease
Pectin	Polysaccharides that, together with hemicelluloses, form the matrix of plant cell walls
Peptides	Combinations of amino acids that make up protein
Pesticide	Toxic chemical product used to kill plant pests (*cf.* herbicide)

Phenotype	The characteristics of an organism that results from the interaction of its genetic constitution with the environment
Phospholipase A2	An enzyme which degrades phospholipids (type A2)
Photobioreactor	Bioreactor dependent on sunlight which is taken up by its content of plant material, usually algae
Photosynthetic efficiency	Efficiency of converting light energy into organic compounds
Phytosanitary	Plant health
Plant cell culture	Growth of tree plant cells or roots of plants in bioreactors
Plant tissue	A section taken from a plant
Plasmid	An extra chromosomal, self-replicating, circular segment of DNA; plasmids (and some viruses) are used as "vectors" for cloning DNA in bacterial "host" cells
Ploidy	Pertaining to chromosome number, e.g. haploid, diploid, polyploid
Polymerase chain reaction (PCR)	The action of an enzyme (polymerase) to produce many copies of a polynucleotide sequence
Polymorphism	The occurrence of different allelic forms of a gene
Polypeptide	A compound that contains many amino acids linked together by peptide bonds
Polysaccharides	High molecular weight polymers, often with functional properties that are industrially useful, e.g. gelling or viscosifying
Polyunsaturates:	Oils containing polyunsaturated fatty acids
Progeny	Offspring
Prolactin	A hormone that stimulates and controls lactation in mammals
Promoter sequence	A regulatory DNA sequence that initiates the expression of a gene
Protein engineering	Generating proteins with subtly modified structures, conferring properties such as higher catalytic specificity or thermal stability

Protein metabolic step	One step in the chain of reactions that take place in a plant and dictate the composition of that plant
Protoplast	A plant cell that has been denuded of its rigid cellulose wall
Protoplast fusion	Fusion of two cells without walls
Protozoa	Microscopic single cell organisms
Pyrethrins	Active constituents of pyrethrum flowers used as insecticides
Recombinant DNA	A DNA molecule formed by joining DNA segments from two or more sources
Recombinant protein	A protein made, usually in bacteria, yeast or other cultured cells as the product of a gene in a recombinant DNA molecule
Restriction enzyme	See restriction nuclease
Restriction fragment length polymorphism (RFLP)	Fragments of differing lengths of DNA that are produced by cutting DNA with restriction enzymes
Restriction nuclease	A bacterial enzyme that cuts DNA at a specific nucleotide sequence
Retroviral vectors	Gene transfer systems based on viruses that have RNA as their genetic material
Rinderpest	Cattle plague; a viral infection of cattle, sheep and goats
RNA	Ribonucleic acid; RNA is similar in structure to DNA. It is a key link in the chain whereby DNA makes RNA make protein
Ruminant farm animals	Animals such as cattle, sheep and goats that possess compound stomachs and regurgitate their food
Safflower	Thistle-like plant yielding an edible oil and a red dye used in oil paints
Saturates	Oils containing saturated fatty acids
Secondary metabolite	Metabolite that is not required by the producing organism for its life-support system
Self-incompatibility	A mechanism in flowering plants that prevents fertilization and development of an embryo following pollination by a genetically identical individual; it prevents inbreeding

Senescence	The advanced phase of the ageing process of an organism
Serum albumin	A globular protein obtained from blood and body fluids
Sexed embryos	Embryos separated according to sex
Single-cell protein	Cells, or protein extracts, of micro-organisms grown in large quantities for use as human or animal protein supplements
Sires	Male animals used for breeding
Somatic cell	Any of the cells of an organism, except cells that give rise to sperm and eggs
Spore	A dormant cellular form, derived from a bacterial or a fungal cell, that is devoid of metabolic activity and that can give rise to a vegetative cell upon germination; it is dehydrated and can survive for prolonged periods of time under drastic environmental conditions
Stem cell	An undifferentiated cell which divides giving rise, usually by a succession of stages, to a mature functional cell
Strain	A variant group within a species
Symbiotic association	An intimate partnership between two organisms, in which the mutual advantages normally outweigh the disadvantages
Tetraploid	Possessing four sets of chromosomes
Thermophilic bacteria	Bacteria that thrive at high temperatures
Toxin	A chemical produced by a pathogen that causes damage to host cell
Transgenic animals or plants	Animals or plants where hereditary DNA has been augmented by the addition of DNA from a source other than parental germplasm
Trichinosis	Infestation with the nematode worm Trichina
Triploid	A cell or organism containing three times the haploid number of chromosomes
Triticale	Cross between wheat and rye
Trypsin inhibitors	Substances inactivating the enzyme trypsin which is needed for digestion of peptides

Vector	An agent of transmission; for example a DNA vector is a self-replicating molecule of DNA that transmits genetic information from one cell or organism to another. Plasmids (and some viruses) are used as "vectors" for DNA in bacterial cloning
Vernalisation	The cold treatment of partially germinated seeds
Viral coat protein	Protein present in the outer layer of a virus
Viral pathogen	Disease-causing virus
Virus	An extremely small infectious agent
X-ray crystallography	Determination of crystal structure from X-ray diffraction patterns

ALSO AVAILABLE

Biotechnology and Patent Protection : An International Review *by F.K. Beier, R.S. Crespi, J. Straus* (1985)
(93 85 05 1) ISBN 92-64-12757-7   FF100   £14.00   US$25.00   DM42
Biotechnology and the Changing Role of Government (1988)
(93 88 04 1) ISBN 92-64-13072-1   FF90   £11.00   US$20.00   DM39
Biotechnology: Economic and Wider Impacts (1989)
(93 89 01 1) ISBN 92-64-13196-5   FF95   £11.50   US$20.00   DM39
Safety Considerations for Biotechnology (1992)
(93 91 05 1) ISBN 92-64-13641-X   FF70   £10.00   US$18.50   DM29

Prices charged at the OECD Bookshop.
THE OECD CATALOGUE OF PUBLICATIONS and supplements will be sent free of charge
on request addressed either to OECD Publications Service,
or to the OECD Distributor in your country.

MAIN SALES OUTLETS OF OECD PUBLICATIONS
PRINCIPAUX POINTS DE VENTE DES PUBLICATIONS DE L'OCDE

ARGENTINA – ARGENTINE
Carlos Hirsch S.R.L.
Galería Güemes, Florida 165, 4° Piso
1333 Buenos Aires   Tel. (1) 331.1787 y 331.2391
  Telefax: (1) 331.1787

AUSTRALIA – AUSTRALIE
D.A. Book (Aust.) Pty. Ltd.
648 Whitehorse Road, P.O.B 163
Mitcham, Victoria 3132   Tel. (03) 873.4411
  Telefax: (03) 873.5679

AUSTRIA – AUTRICHE
Gerold & Co.
Graben 31
Wien I   Tel. (0222) 533.50.14

BELGIUM – BELGIQUE
Jean De Lannoy
Avenue du Roi 202
B-1060 Bruxelles   Tel. (02) 538.51.69/538.08.41
  Telefax: (02) 538.08.41

CANADA
Renouf Publishing Company Ltd.
1294 Algoma Road
Ottawa, ON K1B 3W8   Tel. (613) 741.4333
  Telefax: (613) 741.5439
Stores:
61 Sparks Street
Ottawa, ON K1P 5R1   Tel. (613) 238.8985
211 Yonge Street
Toronto, ON M5B 1M4   Tel. (416) 363.3171
Les Éditions La Liberté Inc.
3020 Chemin Sainte-Foy
Sainte-Foy, PQ G1X 3V6   Tel. (418) 658.3763
  Telefax: (418) 658.3763

Federal Publications
165 University Avenue
Toronto, ON M5H 3B8   Tel. (416) 581.1552
  Telefax: (416) 581.1743

CHINA – CHINE
China National Publications Import
Export Corporation (CNPIEC)
P.O. Box 88
Beijing   Tel. 403.5533
  Telefax: 401.5664

DENMARK – DANEMARK
Munksgaard Export and Subscription Service
35, Nørre Søgade, P.O. Box 2148
DK-1016 København K   Tel. (33) 12.85.70
  Telefax: (33) 12.93.87

FINLAND – FINLANDE
Akateeminen Kirjakauppa
Keskuskatu 1, P.O. Box 128
00100 Helsinki   Tel. (358 0) 12141
  Telefax: (358 0) 121.4441

FRANCE
OECD/OCDE
Mail Orders/Commandes par correspondance:
2, rue André-Pascal
75775 Paris Cedex 16   Tel. (33-1) 45.24.82.00
Telefax: (33-1) 45.24.85.00   or (33-1) 45.24.81.76
  Telex: 620 160 OCDE

OECD Bookshop/Librairie de l'OCDE :
33, rue Octave-Feuillet
75016 Paris   Tel. (33-1) 45.24.81.67
  (33-1) 45.24.81.81

Documentation Française
29, quai Voltaire
75007 Paris   Tel. 40.15.70.00

Gibert Jeune (Droit-Économie)
6, place Saint-Michel
75006 Paris   Tel. 43.25.91.19

Librairie du Commerce International
10, avenue d'Iéna
75016 Paris   Tel. 40.73.34.60

Librairie Dunod
Université Paris-Dauphine
Place du Maréchal de Lattre de Tassigny
75016 Paris   Tel. 47.27.18.56

Librairie Lavoisier
11, rue Lavoisier
75008 Paris   Tel. 42.65.39.95

Librairie L.G.D.J. - Montchrestien
20, rue Soufflot
75005 Paris   Tel. 46.33.89.85

Librairie des Sciences Politiques
30, rue Saint-Guillaume
75007 Paris   Tel. 45.48.36.02

P.U.F.
49, boulevard Saint-Michel
75005 Paris   Tel. 43.25.83.40

Librairie de l'Université
12a, rue Nazareth
13100 Aix-en-Provence   Tel. (16) 42.26.18.08

Documentation Française
165, rue Garibaldi
69003 Lyon   Tel. (16) 78.63.32.23

GERMANY – ALLEMAGNE
OECD Publications and Information Centre
Schedestrasse 7
D-W 5300 Bonn 1   Tel. (0228) 21.60.45
  Telefax: (0228) 26.11.04

GREECE – GRÈCE
Librairie Kauffmann
Mavrokordatou 9
106 78 Athens   Tel. 322.21.60
  Telefax: 363.39.67

HONG-KONG
Swindon Book Co. Ltd.
13–15 Lock Road
Kowloon, Hong Kong   Tel. 366.80.31
  Telefax: 739.49.75

ICELAND – ISLANDE
Mál Mog Menning
Laugavegi 18, Pósthólf 392
121 Reykjavik   Tel. 162.35.23

INDIA – INDE
Oxford Book and Stationery Co.
Scindia House
New Delhi 110001   Tel.(11) 331.5896/5308
  Telefax: (11) 332.5993
17 Park Street
Calcutta 700016   Tel. 240832

INDONESIA – INDONÉSIE
Pdii-Lipi
P.O. Box 4298
Jakarta 12042   Tel. 583467
  Telex: 62 875

IRELAND – IRLANDE
TDC Publishers – Library Suppliers
12 North Frederick Street
Dublin 1   Tel. 74.48.35/74.96.77
  Telefax: 74.84.16

ISRAEL
Electronic Publications only
Publications électroniques seulement
Sophist Systems Ltd.
71 Allenby Street
Tel-Aviv 65134   Tel. 3-29.00.21
  Telefax: 3-29.92.39

ITALY – ITALIE
Libreria Commissionaria Sansoni
Via Duca di Calabria 1/1
50125 Firenze   Tel. (055) 64.54.15
  Telefax: (055) 64.12.57
Via Bartolini 29
20155 Milano   Tel. (02) 36.50.83

Editrice e Libreria Herder
Piazza Montecitorio 120
00186 Roma   Tel. 679.46.28
  Telefax: 678.47.51

Libreria Hoepli
Via Hoepli 5
20121 Milano   Tel. (02) 86.54.46
  Telefax: (02) 805.28.86

Libreria Scientifica
Dott. Lucio de Biasio 'Aeiou'
Via Coronelli, 6
20146 Milano   Tel. (02) 48.95.45.52
  Telefax: (02) 48.95.45.48

JAPAN – JAPON
OECD Publications and Information Centre
Landic Akasaka Building
2-3-4 Akasaka, Minato-ku
Tokyo 107   Tel. (81.3) 3586.2016
  Telefax: (81.3) 3584.7929

KOREA – CORÉE
Kyobo Book Centre Co. Ltd.
P.O. Box 1658, Kwang Hwa Moon
Seoul   Tel. 730.78.91
  Telefax: 735.00.30

MALAYSIA – MALAISIE
Co-operative Bookshop Ltd.
University of Malaya
P.O. Box 1127, Jalan Pantai Baru
59700 Kuala Lumpur
Malaysia   Tel. 756.5000/756.5425
  Telefax: 755.4424

NETHERLANDS – PAYS-BAS
SDU Uitgeverij
Christoffel Plantijnstraat 2
Postbus 20014
2500 EA 's-Gravenhage   Tel. (070 3) 78.99.11
Voor bestellingen: Tel. (070 3) 78.98.80
  Telefax: (070 3) 47.63.51

NEW ZEALAND
NOUVELLE-ZÉLANDE
Legislation Services
P.O. Box 12418
Thorndon, Wellington   Tel. (04) 496.5652
  Telefax: (04) 496.5698

NORWAY – NORVÈGE
Narvesen Info Center – NIC
Bertrand Narvesens vei 2
P.O. Box 6125 Etterstad
0602 Oslo 6   Tel. (02) 57.33.00
  Telefax: (02) 68.19.01

PAKISTAN
Mirza Book Agency
65 Shahrah Quaid-E-Azam
Lahore 3   Tel. 66.839
  Telex: 44886 UBL PK. Attn: MIRZA BK

PORTUGAL
Livraria Portugal
Rua do Carmo 70-74
Apart. 2681
1117 Lisboa Codex   Tel.: (01) 347.49.82/3/4/5
  Telefax: (01) 347.02.64

SINGAPORE – SINGAPOUR
Information Publications Pte
Golden Wheel Bldg.
41, Kallang Pudding, #04-03
Singapore 1334
Tel. 741.5166
Telefax: 742.9356

SPAIN – ESPAGNE
Mundi-Prensa Libros S.A.
Castelló 37, Apartado 1223
Madrid 28001
Tel. (91) 431.33.99
Telefax: (91) 575.39.98

Librería Internacional AEDOS
Consejo de Ciento 391
08009 – Barcelona
Tel. (93) 488.34.92
Telefax: (93) 487.76.59
Llibreria de la Generalitat
Palau Moja
Rambla dels Estudis, 118
08002 – Barcelona
(Subscripcions) Tel. (93) 318.80.12
(Publicacions) Tel. (93) 302.67.23
Telefax: (93) 412.18.54

SRI LANKA
Centre for Policy Research
c/o Colombo Agencies Ltd.
No. 300-304, Galle Road
Colombo 3
Tel. (1) 574240, 573551-2
Telefax: (1) 575394, 510711

SWEDEN – SUÈDE
Fritzes Fackboksföretaget
Box 16356
Regeringsgatan 12
103 27 Stockholm
Tel. (08) 23.89.00
Telefax: (08) 20.50.21
Subscription Agency-Agence d'abonnements
Wennergren-Williams AB
Nordenflychtsvägen 74
Box 30004
104 25 Stockholm
Tel. (08) 13.67.00
Telefax: (08) 618.62.32

SWITZERLAND – SUISSE
Maditec S.A. (Books and Periodicals - Livres
et périodiques)
Chemin des Palettes 4
1020 Renens/Lausanne
Tel. (021) 635.08.65
Telefax: (021) 635.07.80

Mail orders only - Commandes
par correspondance seulement
Librairie Payot
C.P. 3212
1002 Lausanne
Telefax: (021) 311.13.92

Librairie Unilivres
6, rue de Candolle
1205 Genève
Tel. (022) 320.26.23
Telefax: (022) 329.73.18

Subscription Agency - Agence d'abonnement
Naville S.A.
38 avenue Vibert
1227 Carouge
Tél.: (022) 308.05.56/57
Telefax: (022) 308.05.88

See also – Voir aussi :
OECD Publications and Information Centre
Schedestrasse 7
D-W 5300 Bonn 1 (Germany)
Tel. (49.228) 21.60.45
Telefax: (49.228) 26.11.04

TAIWAN – FORMOSE
Good Faith Worldwide Int'l. Co. Ltd.
9th Floor, No. 118, Sec. 2
Chung Hsiao E. Road
Taipei
Tel. (02) 391.7396/391.7397
Telefax: (02) 394.9176

THAILAND – THAÏLANDE
Suksit Siam Co. Ltd.
113, 115 Fuang Nakhon Rd.
Opp. Wat Rajbopith
Bangkok 10200
Tel. (662) 251.1630
Telefax: (662) 236.7783

TURKEY – TURQUIE
Kültur Yayinlari Is-Türk Ltd. Sti.
Atatürk Bulvari No. 191/Kat. 13
Kavaklidere/Ankara
Tel. 428.11.40 Ext. 2458
Dolmabahce Cad. No. 29
Besiktas/Istanbul
Tel. 160.71.88
Telex: 43482B

UNITED KINGDOM – ROYAUME-UNI
HMSO
Gen. enquiries
Tel. (071) 873 0011
Postal orders only:
P.O. Box 276, London SW8 5DT
Personal Callers HMSO Bookshop
49 High Holborn, London WC1V 6HB
Telefax: (071) 873 8200
Branches at: Belfast, Birmingham, Bristol, Edin-
burgh, Manchester

UNITED STATES – ÉTATS-UNIS
OECD Publications and Information Centre
2001 L Street N.W., Suite 700
Washington, D.C. 20036-4910 Tel. (202) 785.6323
Telefax: (202) 785.0350

VENEZUELA
Libreria del Este
Avda F. Miranda 52, Aptdo. 60337
Edificio Galipán
Caracas 106
Tel. 951.1705/951.2307/951.1297
Telegram: Libreste Caracas

YUGOSLAVIA – YOUGOSLAVIE
Jugoslovenska Knjiga
Knez Mihajlova 2, P.O. Box 36
Beograd
Tel. (011) 621.992
Telefax: (011) 625.970

Orders and inquiries from countries where Distribu-
tors have not yet been appointed should be sent to:
OECD Publications Service, 2 rue André-Pascal,
75775 Paris Cedex 16, France.

Les commandes provenant de pays où l'OCDE n'a
pas encore désigné de distributeur devraient être
adressées à : OCDE, Service des Publications,
2, rue André-Pascal, 75775 Paris Cedex 16, France.

Subscription to OECD periodicals may also be
placed through main subscription agencies.

Les abonnements aux publications périodiques de
l'OCDE peuvent être souscrits auprès des
principales agences d'abonnement.

OECD PUBLICATIONS, 2 rue André-Pascal, 75775 PARIS CEDEX 16
PRINTED IN FRANCE
(93 92 03 1) ISBN 92-64-13725-4 - No. 46047 1992

5104